Springer Theses

Recognizing Outstanding Ph.D. Research

Aims and Scope

The series "Springer Theses" brings together a selection of the very best Ph.D. theses from around the world and across the physical sciences. Nominated and endorsed by two recognized specialists, each published volume has been selected for its scientific excellence and the high impact of its contents for the pertinent field of research. For greater accessibility to nonspecialists, the published versions include an extended introduction, as well as a foreword by the student's supervisor explaining the special relevance of the work for the field. As a whole, the series will provide a valuable resource both for newcomers to the research fields described, and for other scientists seeking detailed background information on special questions. Finally, it provides an accredited documentation of the valuable contributions made by today's younger generation of scientists.

Theses are accepted into the series by invited nomination only and must fulfill all of the following criteria

- They must be written in good English.
- The topic should fall within the confines of Chemistry, Physics, Earth Sciences, Engineering and related interdisciplinary fields such as Materials, Nanoscience, Chemical Engineering, Complex Systems and Biophysics.
- The work reported in the thesis must represent a significant scientific advance.
- If the thesis includes previously published material, permission to reproduce this must be gained from the respective copyright holder.
- They must have been examined and passed during the 12 months prior to nomination.
- Each thesis should include a foreword by the supervisor outlining the significance of its content.
- The theses should have a clearly defined structure including an introduction accessible to scientists not expert in that particular field.

More information about this series at http://www.springer.com/series/8790

Wei Chen

Explosive Percolation in Random Networks

Doctoral Thesis accepted by
Peking University, Beijing, China

Author
Dr. Wei Chen
Peking University
Beijing
China

Supervisor
Prof. Zhiming Zheng
LMIB and School of Mathematics
and Systems Sciences
Beihang University
Beijing
China

ISSN 2190-5053 ISSN 2190-5061 (electronic)
ISBN 978-3-662-51539-6 ISBN 978-3-662-43739-1 (eBook)
DOI 10.1007/978-3-662-43739-1

Mathematics Subject Classification: 60K35, 65C05, 65C20

Springer Heidelberg New York Dordrecht London

Printed on acid-free paper

Springer is part of Springer Science+Business Media (www.springer.com)

It is not enough to be in the right place at the right time. You should also have an open mind at the right time

Paul Erdös

Supervisor's Foreword

Percolation transition, the transition from a disconnected state to a connected one, has been regarded as a fundamental model of phase transitions in nonequilibrium systems. One of the most fundamental characteristics of a phase transition is its order, i.e., whether the macroscopic quantity affects changes continuously or discontinuously at the transition point. The percolation phase transitions in random networks were originally considered to be robust continuous phase transitions. A few years ago, it was reported that random networks under the Achlioptas process underwent a discontinuous transition that was called "explosive percolation." It was finally proven that the transition is actually continuous in the thermodynamic limit, although it is extremely abrupt. Thus, understanding what types of phenomena can lead to discontinuous phase transitions in the connectivity of random networks is an outstanding challenge.

This thesis contains the results of Dr. Wei Chen's research when he was working on his Ph.D., with the aim of understanding the BFW model that was first introduced by Tom Bohman, Alan Frieze, and Nicholas Charles Wormald, a typical model that exhibits a discontinuous percolation transition. The principal results dealt with the analysis of several important and unusual behaviors observed in this model, including discontinuous percolation transition, the underlying mechanism of discontinuous transition, multiple discontinuous transitions, and unstable giant components.

The layout of the thesis is as follows: In Chap. 1, the author recalls the development of network science, the theory of percolation, and explosive percolation in random networks. In Chap. 2, he studies the nature of percolation transition and number of giant components in the Bohman–Frieze–Wormald model. Chapter 3 is devoted to the study of the underlying mechanism of discontinuous transition in this model. In Chap. 4, the author shows that at some point in the supercritical regime, the fraction of nodes in the largest component stops growing and eventually a second giant component emerges in a continuous percolation transition. The author also establishes many features of the second continuous percolation that include scaling exponents and relations. In the final chapter, he investigates whether the location of the largest jump coincides with the percolation threshold for a range of processes, such as Erdös–Rényi percolation, and percolation via edge competition and via growth by overtaking.

The results arrived at, in this thesis, were mainly obtained through numerical simulations. Therefore, there is still great potential for further theoretical analysis in order to understand these results in a rigorous way. The methods developed in carrying out this work are expected to be useful for analyzing other models of percolation. It is also hoped that the thesis can serve as a valuable reference for current development in problems related to both the theory and application of percolation.

Beijing, April 2014 Prof. Zhiming Zheng

Acknowledgments

First, I would like to express my deep gratitude to my Ph.D. supervisor, Prof. Zhiming Zheng, for offering me systematic guidance, excellent teaching, and encouragement throughout my 5 years of graduate studies.

Second, I am very grateful to Prof. Raissa M. D'Souza, who not only provided me with the topic for my Ph.D. thesis, but also gave me many insightful suggestions and valuable instruction in research when I was at the University of California, Davis, for a year-and-a-half. I am also grateful for her financial support during this period.

Finally, I would also give my sincere thanks to Profs. Wolfgang Polonik, Naoki Saito, Guanrong Chen, Choy Heng Lai, Bican Xia, Ke Xu, and Drs. Ning Ning Chung, Xin Jiang, Zhanli Zhang, Lili Ma, Shaoting Tang for their help with my research.

This book was partially supported by the National High Technology Research and Development Program of China (the 863 program), under grant No. 2014AA015103, the National Basic Research Program of China (the 973 Program), under grant No. 2013CB329602, and the National Natural Science Foundation of China (NSFC), under grant Nos. 61232010 and 11305219.

Contents

Abstract

Random networks, a collection of nodes and random edges, provide a useful abstraction of the structure of many complex systems, from biological systems and social systems to technological systems and economical systems. Viewed in a random network setting, once the density of edges exceeds a critical threshold, the system undergoes a sudden transition from small, scattered components to global connectivity, where the size of the largest connected component transitions from microscopic to macroscopic in size; mathematicians and physicists call this "percolation." Percolation governs the dynamics of many social and physical systems as well as in the epidemic spreading of infectious diseases and information propagation. The Erdös–Rényi Random Network was considered robust continuous transition at the percolation threshold. An important and interesting question is whether the transition can be discontinuous or "explosive" under some special rules of connecting edges. In this work, we study a model of percolation that was first proposed by T. Bohman, A. Frieze, and N.C. Wormald (see Random Struct. Algorithms, 25, 432 (2004)). We find that this model not only exhibits discontinuous transition at the percolation threshold, but also many other unusual critical behaviors that can trigger much interesting follow on work. The description of the work is divided into four parts as follows:

We begin Chap. 2 with our study of the nature of percolation transition and the number of giant components in the Bohman–Frieze–Wormald model. Starting from a collection of "n" isolated nodes, potential edges chosen uniformly at random from the complete graph are examined one at a time, while a cap, "k," on the maximum allowed component size is enforced. Edges whose addition would exceed k can be simply rejected, provided the accepted fraction of edges never becomes smaller than a function that decreases with k, such as $g(k) = \alpha + (2k)^{-1/2}$. We show that multiple giant components appear simultaneously in a strongly discontinuous percolation transition and remain distinct in the supercritical regime. Furthermore, tuning the value of α determines the number of such components with smaller α leading to an increasingly delayed and more explosive transition. The location of the critical point and strongly discontinuous nature is not affected if only edges that span components are sampled.

In Chap. 3, we study the underlying mechanism of discontinuous transition in the Bohman–Frieze–Wormald model. This mechanism is the domination of overtaking in the growth of the largest component. Here, we assume that $g(k) = 1/2 + (2k)^{-\beta}$. We show that if $\beta < 1$, it is always possible to reject a sampled edge, and the growth in the largest component is dominated by an overtaking mechanism leading to a discontinuous transition. If $\beta > 1$, once $k \geq n^{1/\beta}$, there are situations when a sampled edge must be accepted, leading to direct growth dominated by stochastic fluctuations and a "weakly" discontinuous transition. We also show that the distribution of component sizes and the evolution of component sizes are distinct from those previously observed and show no finite size effects for the range of β studied.

In Chap. 4, we focus on the regime $\alpha \in [0.6, 0.95]$, where it is known that only one giant component initially appears at the discontinuous phase transition. We show that at some point in the supercritical regime, the fraction of nodes in the largest component, denoted as C_1, stops growing and eventually a second giant component emerges in a continuous percolation transition. The delay between the emergence of two giant components and their asymptotic sizes both depends on the value of α and we establish, by several techniques, that there exists a bifurcation point α_c. For $\alpha \in [0.6, \alpha_c)$, C_1 stops growing the instant it emerges and the delay between the emergence of two giant components decreases with increasing α. For $\alpha \in (\alpha_c, 0.95]$, in contrast, C_1 continues growing into the supercritical regime and the delay between the emergence of two giant components increases with increasing α. As we show, α_c marks the minimal delay possible between the emergence of two giant components. We also establish many features of the continuous percolation of C_2, including scaling exponents and relations.

The location and nature of the percolation transition in random networks is a subject of intense interest. A series of graph evolution processes has been introduced and leads to discontinuous percolation transitions where the addition of a single edge causes the size of the largest component to exhibit a significant macroscopic jump in the thermodynamic limit. These processes can have additional exotic behaviors, such as displaying a "Devil's staircase" of discrete jumps in the supercritical regime. In Chap. 5, we investigate whether the location of the largest jump coincides with the percolation threshold for a range of processes, such as Erdös–Rényi percolation, percolation via edge competition, and via growth by overtaking. We find that the largest jump asymptotically occurs at the percolation transition for Erdös–Rényi and other processes exhibiting global continuity, including models exhibiting an "explosive" transition. However, for percolation processes exhibiting genuine discontinuities, the behavior is substantially richer. In percolation models where the order parameter exhibits a staircase, the largest discontinuity generically does not coincide with the percolation transition. For the generalized Bohman–Frieze–Wormald model, it depends on the model parameter. Distinct parameter regimes well in the supercritical regime feature unstable discontinuous transitions—a novel and unexpected phenomenon in percolation. We thus show that seemingly and genuinely discontinuous percolation transitions can involve a rich behavior in supercriticality, a regime that has been largely ignored in percolation.

Keywords Community · Complex networks · Explosive percolation · Giant connected component · Random graph · Phase transition

Chapter 1
Introduction

1.1 Network Science: An Emerging Discipline

Networks, or graphs (a collection of nodes with edges connecting them), have long been studied in a prolific branch of mathematics known as "graph theory" and are often used by scientists to model the structure of many complex systems, ranging from technological systems [1–3] and biological systems [4, 5] to social systems [6, 7] and economic systems [8, 9]. Human beings are surrounded by a variety of different types of networks. Examples include networks of scientific collaborations [10, 11], in which nodes are scientists and edges connecting co-authors; the Internet [2, 3], in which nodes represent the computers or routers and edges represent cables used to transmit data; the protein homology network [12], in which the nodes are the proteins and weighted edges represent the degree of similarity between two proteins; and food webs in ecosystems [5], in which nodes are species and edges show the predator and prey interactions.

In the middle of the last century, due to the absence of reliable data about complex systems, the connections between nodes were considered to be random, and networks were mostly modeled by random graphs. The most classical and widely used random graph model is the Erdös-Rényi Random Graph, which was first proposed by the Hungarian mathematicians Paul Erdös and Alfred Rényi in the middle of the twentieth century [13, 14]. In this model, every pair of nodes is connected with a certain probability p in a network with a fixed number of nodes, resulting in a Poisson degree distribution in the network. At the end of the last century, with the rapid development of complex systems (including the Internet and the World Wide Web), large-scale information about the structure of complex systems became attainable. To the surprise of many, the networks of many complex systems are not Erdös-Rényi random graphs. In 1999, Albert-Laszlo Barabási from the University of Notre Dame, published a paper entitled "Emergence of Scaling in Random Networks" in *Science* [15]. He discovered that the network of the Internet appears to follow a power-law distribution, not a Poisson distribution. These types of networks are also called "scale-free" networks. This discovery indicates that the wiring of the Internet is dominated

W. Chen, *Explosive Percolation in Random Networks*, Springer Theses,
DOI: 10.1007/978-3-662-43739-1_1

by several highly connected hubs, which are not a feature of an Erdös-Rényi random graph. One year before this study, Duncan Watts from Columbia University and Steven Strogatz from Cornell University made another important discovery in their paper "Collective Dynamics of Small-World Networks," which was published in *Nature* [16]. They showed that many networks in nature, such as the brain of the worm *C. elegans*, as well as the network of movie actors in Hollywood and the network of power lines in western America, simultaneously have a small node separation and display a high degree of clustering. These networks are called "small world," which was first proposed in 1967 by Stanley Milgram, a sociologist at Harvard University, [17]. These two papers are considered to be the most distinguished studies in modern network science because they launched a flood of subsequent papers in this field over the past 15 years.

1.2 Percolation in Networks

Percolation is a pervasive concept that arises in atomic and molecular solids in physics as well as in social, technological and natural systems [18–20] such as epidemic or rumor spread [21, 22], information diffusion in online social networks [23], porous media [24], resistor networks [25], forest fires [26] etc. Percolation has been studied to a great extent for several decades because its critical information on the large-scale connectivity of a system can be used to quantify its ability to efficiently transfer information, resources, energy, etc. [27, 28]. A typical percolation model is random networks, in which a phase transition from small, scattered components, to large-scale connectedness, occurs upon the sequential addition of random edges between nodes. Perhaps the most classical and widely used model of random network is the Erdös-Rényi Random Graph. This model starts with N isolated nodes. At every step of the way?, two nodes are randomly selected and connected with an edge. Once the density of edges (number of edges per node in the system) exceeds a critical threshold $p_c = 1/2$, the order of magnitude in the size of the largest connected component suddenly changes from $o(N)$ to $\mathcal{O}(N)$. Mathematical frameworks characterizing the critical properties of the Erdös-Rényi model can be seen in Refs. [13] and [14].

Since the emergence of complex networks at the end of the last century, percolation has been studied in a variety of different types of networks. A random network with (without) pair correlations between nearest-neighbor degrees is called a "correlated (uncorrelated) network". The mathematical algorithm for organizing an arbitrary uncorrelated network, called a "configuration model," was proposed by Michael Molloy and Bruce Reed [29, 30]. Mark Newman developed the generating function technique to provide the analytical solution of the size of the largest connected component and percolation threshold for uncorrelated networks with arbitrary degree distributions [31]. Reuven Cohen et al. first considered the percolation problem of an uncorrelated network under random removal or damage nodes or edges [32]. Duncan Callaway et al. considered a more general percolation problem in uncorrelated scale-free networks where the probability of removing a node depends on its degree. It was discovered that giant connected components in a scale-free network are robust to random attacks but rather vulnerable to intentional damage of highly connected hubs [33, 34]. Percolation on degree-degree correlated networks is further studied, based on generating function technique in Refs. [35–37]. Sergey Dorogovtsev et al. [38], Nehemia Schwartz et al. [39] and M. Angeles Serrano et al. [40] studied the structure of the giant connected component in uncorrelated directed networks. Marian Boguñá and M. Angeles Serrano generalized their theory to uncorrelated networks containing both directed and undirected edges [41]. Duncan Callaway and his group studied percolation phase transition in a simple model of a growing network [42]. Mark Newman and Duncan Watts investigated the site percolation on small-world networks as a simple model of disease propagation and they derived an approximate expression for the percolation probability at which a giant component first emerges [43]. M. Angeles Serrano et al. studied the percolation in self-similar networks and proved that graphs in a general class of self-similar networks have a zero percolation threshold [44]. Hon Wai Lau et al. studied the agglomerative percolation on bipartite networks and found the nonuniversal behavior at the percolation threshold resulting from spontaneous symmetry breaking [45]. James Gleeson et al. [46, 47] and Mark Newman [48] studied the percolation in clustered networks. Generalizations of percolation in networks include k-clique percolation [49, 50] and bootstrap percolation (k-core percolation) [51, 52], Ising models on networks [53, 54], Potts models on networks [55], dynamic models of epidemic spreading [21, 56–60], dynamic models of information diffusion on social networks [61–64], and others.

1.3 Explosive Percolation: A Continuous or Discontinuous Transition?

One of the most fundamental characteristics of a phase transition is its order, i.e., whether the macroscopic quantity it affects changes continuously or discontinuously at the transition point. Continuous transitions are called "second-order" and include many magnetization phenomena, whereas discontinuous transitions are called "first-order," an example of which is the discontinuous drop in entropy when liquid water turns into solid ice at zero degrees.

The percolation phase transitions in random networks were originally considered to be a robust continuous phase transition, such as the Erdös-Rényi model. However, in 2009, Dimitris Achlioptas et al. reported that a variant of the Erdös-Rényi model, which was called the "Achlioptas Process," can undergo a discontinuous transition or "explosive" percolation, as it was called in their paper [65]. In an Achlioptas Process, multiple candidate edges are considered randomly from the complete graph generated by all the nodes, and only one edge is connected to the graph according to a pre-set criterion, while the others are discarded. They showed that if the pre-set criterion is the product rule or sum rule, i.e., choosing the edge connecting two components with a smaller product or sum of their sizes, the transition is discontinuous. The reason given by them is that in a network of size N, only the number of edges added to the graph is of order $o(N)$ during the size of the largest connected component augmenting from $\mathcal{O}(N^\beta)$ to γN for some $0 < \beta, \gamma < 1$. Several months later, explosive percolation transition was further observed in two-dimensional square lattices under the product rule [66, 67], scale-free networks under the product rule [68, 69], random percolation under triangle rule and adjacent rule [70], product rule in human protein homology networks [12], nanotube-based systems [71], and even large-scale social networks [72]. Their results are all based on numerical simulations without strict mathematical proofs.

Is explosive percolation really discontinuous in the thermodynamic limit? Two years later, Rui da Costa et al. introduced mean-field equations to describe the evolution of a special "Achlioptas Process" [73, 74]. They proved, by numerical and analytical approaches, that the product rule actually exhibits a continuous, second-order phase transition, rather than "discontinuous" transition which is generally convinced but with a small critical exponent of the percolation component size. Peter Grassberger studied more Achlioptas Processes and reported that they all exhibit continuous transitions although they belong to different universality classes [75]. Later on, scientists performed extensive numerical simulations with finite-size scaling analysis at criticality to show that more Achlioptas Processes in random graph and two-dimensional square lattices are continuous [76–80]. Terry Riordan and Scott Warnke first applied the probability theory to prove that all Achlioptas Processes in random graphs are continuous [81, 82].

The debate in the nature of explosive percolation gives rise to an outstanding challenge: what are the essential ingredients that lead to discontinuous phase transitions in the connectivity of random networks? Actually, this question was in the spotlight soon after the emergence of explosive percolation. Eric Friedman and Adam Landsberg first proposed the concept of the powder keg, which is defined as the nodes in components of a size larger than N^α, when the size of the largest component is $N^\beta (1 > \alpha > \beta > 0)$ [83]. They found that the fraction of nodes in the powder keg converges to some positive constant as $N \to \infty$ in the subcritical regime, which induces the abrupt discontinuous transition at percolation threshold. Similar to the idea of the powder keg, Hans Hooybergs and Bert von Schaeybroeck proposed an equivalent condition for discontinuous explosive percolation, which is the mean number of nodes per cluster diverges in the thermodynamic limit prior to the transition point. André Moreira et al. suggest that two conditions are sufficient to generate the discontinuous explosive percolation: (1) the size of all growing components should be kept approximately the same, and (2) edges that merge two components should be selected much more frequently than edges inside the components [84]. Young Sul Cho and Byungnam Kahng show that when the dynamic rules of connecting edges suppress the growth of all components, the transition is discontinuous [85]. Other mechanisms that lead to discontinuous transition include cooperative phenomena [86], hierarchical structures [87], correlated percolation [88], restricted Erdös-Rényi models [89], algorithms that explicitly suppress growth of the largest component [90–92], cascading failure in interdependent networks [93], and others. The recent advances in explosive percolation can be found in a review paper by Nikolaos Bastas et al. [94].

In this book, we study a typical explosive percolation model that exhibits a discontinuous transition—Bohman-Frieze-Wormald model. We introduce the mechanism that leads to discontinuous transition [92, 95, 96] and other interesting critical and supercritical behaviors observed in some generalizations of this model, including multiple giant components [92], multiple phase transitions [97–100] and unstable giant components [101].

References

1. Newman, M.E.J.: The structure and function of complex networks. SIAM Rev. **45**(2), 167–256 (2003)
2. Pastor-Satorras, R., Vespignani, A.: Evolution and Structure of the Internet. Cambridge University Press, Cambridge (2004)
3. Faloutsos, M., Faloutsos, P., Faloutsos, C.: On power-law relationships of the internet topology. Comput. Commun. Rev. **29**, 251262 (1999)
4. Girvan, M., Newman, M.E.J.: Community structure in social and biological networks. Proc. Natl. Acad. Sci. USA **99**(12), 7821–7826 (2002)
5. Pascual, M., Dunne, J. A.: (eds) Ecological networks: linking structure to dynamics in food webs. Oxford University Press, Oxford (2006)
6. Wasserman, S., Faust, K.: Social Network Analysis. Cambridge University Press, Cambridge (1994)
7. Scott, J.: Social Network Analysis: A Handbook, 2nd edn. Sage, London (2000)
8. Jackson, M.O., Wolinsky, A.: A strategic model of social and economic networks. J. Econ. Theory **71**(1), 44–74 (1996)
9. Schweitzer, F., Fagiolo, G., Sornette, D., Vega-Redondo, F., Vespignani, A., White, D.R.: Economic networks: the new challenges. Science **325**, 422 (2009)
10. Newman, M.E.J.: The structure of scientific collaboration networks. Proc. Natl. Acad. Sci. USA **98**(2), 404–409 (2001)
11. Barabási, A.-L., Jeong, H., Néda, Z., Ravasz, E., Schubert, A., Vicsek, T.: Evolution of the social network of scientific collaborations. Phys. A **311**, 590 (2002)
12. Rozenfeld, H.D., Gallos, L.K., Makse, H.A.: Explosive percolation in the human protein homology network. Eur. Phys. J. B **75**, 305–310 (2010)
13. Erdös, P., Rényi, A.: On random graphs I. Publ. Math. (Debrecen) **6**, 290 (1959)
14. Erdös, P., Rényi, A.: On the evolution of random graphs. Publ. Math. Inst. Hungar. Acad. Sci. **5**, 17 (1960)
15. Barabási, A.-L., Albert, R.: Emergence of scaling in random networks. Science **286**, 509 (1999)
16. Watts, D.J., Strogatz, S.: Collective dynamics of small-world networks. Nature **393**, 440 (1998)
17. Milgram, S.: The small-world problem. Psychol. Today **1**, 61 (1967)
18. Strogatz, S.H.: Exploring complex networks. Nature **410**, 268276 (2001)
19. Ben-Avraham, D., Havlin, S.: Diffusion and Reactions in Fractals and Disordered Systems. Cambridge University Press, Cambridge (2001)
20. Solomon, S.G., Weisbucha, L.D.A., Janc, N., Stauffer, D.: Social percolation model. Phys. A **277**, 239247 (2000)
21. Newman, M.E.J.: Spread of epidemic disease on networks. Phys. Rev. E **66**, 016128 (2002)
22. Moreno, Y., Nekovee, M., Pacheco, A.F.: Dynamics of rumor spreading in complex networks. Phys. Rev. E **69**, 066130 (2004)

23. Barrat, A., Barthelemy, M., Vespignani, A.: Dynamical Processes on Complex Networks. Cambridge University Press, Cambridge (2008)
24. Chandler, R., Koplik, J., Lerman, K., Willemsen, J.F.: Capillary displacement and percolation in porous media. J. Fluid Mech. **119**, 248–267 (1982)
25. Kirkpatrick, S.: Percolation and conduction. Rev. Mod. Phys. **45**, 574–588 (1973)
26. Drossel, B., Schwabl, F.: Self-organized critical forest-fire model. Phys. Rev. Lett. **69**, 1629–1632 (1992)
27. Stauffer, D., Aharony, A.: Introduction to Percolation Theory. Taylor & Francis, London (1994)
28. Sahimi, M.: Applications of Percolation Theory. Taylor & Francis, London (1994)
29. Molloy, M., Reed, B.A.: A critical point for random graphs with a given degree sequence. Random Struct. Algorithms **6**, 161 (1995)
30. Molloy, M., Reed, B.A.: The size of the giant component of a random graph with a given degree sequence. Comb. Probab. Comput. **7**, 295 (1998)
31. Newman, M.E.J., Strogatz, S.H., Watts, D.J.: Random graphs with arbitrary degree distributions and their applications. Phys. Rev. E **64**, 026118 (2001)
32. Cohen, R., Erez, K., Ben-Avraham, D., Havlin, S.: Resilience of the internet to random breakdowns. Phys. Rev. Lett. **85**, 4626 (2000)
33. Cohen, R., Erez, K., Ben-Avraham, D., Havlin, S.: Breakdown of the internet under intentional attack. Phys. Rev. Lett. **86**, 3682 (2001)
34. Albert, R., Jeong, H., Barabsi, A.-L.: Error and attack tolerance of complex networks. Nature **406**, 378 (2000)
35. Newman, M.E.J.: Assortative mixing in networks. Phys. Rev. Lett. **89**, 208701 (2002)
36. Boguñá, M., Pastor-Satorras, R., Vespignani, A.: Epidemic spreading in complex networks with degree correlations. Lect. Notes Phys. **625**, 127 (2003)
37. Vázquez, A., Moreno, Y.: Resilience to damage of graphs with degree correlations. Phys. Rev. E **67**, 015101(R) (2003)
38. Dorogovtsev, S.N., Mendes, J.F.F., Samukhin, A.N.: Giant strongly connected component of directed networks. Phys. Rev. E **64**, 025101(R) (2001)
39. Schwartz, N., Cohen, R., Ben-Avraham, D., Barabási, A.-L., Havlin, S.: Percolation in directed scale-free networks. Phys. Rev. E **66**, 015104 (2002)
40. Serrano, M.Á., De Los Rios, P.: Interfaces and the edge percolation map of random directed networks. Phys. Rev. E **76**, 056121 (2007)
41. Boguñá, M., Serrano, M.A.: Generalized percolation in random directed networks. Phys. Rev. E **72**, 016106 (2005)
42. Callaway, D.S., Hopcroft, J.E., Kleinberg, J.M., Newman, M.E.J., Strogatz, S.H.: Are randomly grown graphs really random? Phys. Rev. E **64**, 041902 (2001)
43. Newman, M.E.J., Watts, D.J.: Scaling and percolation in the small-world network model. Phys. Rev. E **60**, 7332 (1999)
44. Serrano, M.Á., Krioukov, D., Boguñá, M.: Percolation in self-similar networks. Phys. Rev. Lett. **106**, 048701 (2011)
45. Lau, H.W., Paczuski, M., Grassberger, P.: Agglomerative percolation on bipartite networks: nonuniversal behavior due to spontaneous symmetry breaking at the percolation threshold. Phys. Rev. E **86**, 011118 (2012)
46. Gleeson, J.P.: Bond percolation on a class of clustered random networks. Phys. Rev. E **80**, 036107 (2009)
47. Gleeson, J.P., Melnik, S., Hackett, A.: How clustering affects the bond percolation threshold in complex networks. Phys. Rev. E **81**, 066114 (2010)
48. Newman, M.E.J.: Random graphs with clustering. Phys. Rev. Lett. **103**, 058701 (2009)
49. Deréyi, I., Palla, G., Vicsek, T.: Clique percolation in random networks. Phys. Rev. Lett. **94**, 160202 (2005)
50. Palla, G., Deréyi, I., Vicsek, T.: The critical point of k-clique percolation in the Erdös-Rényi graph. J. Stat. Phys. **128**, 219 (2007)

51. Dorogovtsev, S.N., Goltsev, A.V., Mendes, J.F.F.: K-core organization of complex networks. Phys. Rev. Lett. **96**, 040601 (2006)
52. Dorogovtsev, S.N., Goltsev, A.V., Mendes, J.F.F.: Critical phenomena in complex networks. Rev. Mod. Phys. **80**, 1275–1335 (2008)
53. Gitterman, M.: Small-world phenomena in physics: the Ising model. J. Phys. A: Math. Gen. **33**, 8373 (2000)
54. Dorogovtsev, S.N., Goltsev, A.V., Mendes, J.F.F.: Ising model on networks with an arbitrary distribution of connections. Phys. Rev. E **66**, 016104 (2002)
55. Dorogovtsev, S.N., Goltsev, A.V., Mendes, J.F.F.: Potts model on complex networks Eur. Phys. J. B **38**, 177 (2004)
56. Pastor-Satorras, R., Vespignani, A.: Epidemic spreading in scale-free networks. Phys. Rev. Lett. **86**, 3200–3203 (2001)
57. Pastor-Satorras, R., Vespignani, A.: Epidemic dynamics in finite size scale-free networks. Phys. Rev. E **65**, 035108(R) (2002)
58. Boguñá, M., Pastor-Satorras, R., Vespignani, A.: Absence of epidemic threshold in scale-free networks with degree correlations. Phys. Rev. Lett. **90**, 028701 (2003)
59. Ben-Naim, E., Krapivsky, P.L.: Size of outbreaks near the epidemic threshold. Phys. Rev. E **69**, 050901(R) (2004)
60. Moreno, Y., Pastor-Satorras, R., Vespignani, A.: Epidemic outbreaks in complex heterogeneous networks. Eur. Phys. J. B **26**, 521 (2002)
61. Watts, D.J.: A simple model of global cascades on random networks. Proc. Natl. Acad. Sci. USA **99**, 5766–5771 (2002)
62. Newman, M.E.J., Watts, D.J., Strogatz, S.H.: Random graph models of social networks. Proc. Natl. Acad. Sci. USA **99**, 2566–2572 (2002)
63. Kempe, D., Kleinberg, J., Tardos, É: Maximizing the spread of influence through a social network. In: Proceedings of the 9th ACM SIGKDD international conference on knowledge discovery and data mining, pp 137–146 (2003)
64. Gruhl, D., Guha, R., Liben-Nowell, D., Tomkins, A.: Information diffusion through blogspace. In: Proceedings of the 13th international conference on World Wide Web, pp 491–501 (2004)
65. Achlioptas, D.D., 'Souza, R.M., Spencer, J.: Explosive percolation in random networks. Science **323**, 1453–1455 (2009)
66. Ziff, R.M.: Explosive growth in biased dynamic percolation on two-dimensional regular lattice networks. Phys. Rev. Lett. **103**, 045701 (2009)
67. Ziff, R.M.: Scaling behavior of explosive percolation on the square lattice. Phys. Rev. E **82**, 051105 (2010)
68. Radicchi, F., Fortunato, S.: Explosive percolation in scale-free networks. Phys. Rev. Lett. **103**, 168701 (2009)
69. Radicchi, F., Fortunato, S.: Explosive percolation: a numerical analysis. Phys. Rev. E **81**, 036110 (2010)
70. D'Souza, R.M., Mitzenmacher, M.: Local cluster aggregation models of explosive percolation. Phys. Rev. Lett. **104**, 195702 (2010)
71. Kim, Y., Yun, Y., Yook, S.H.: Explosive percolation in a nanotube-based system. Phys. Rev. E **82**, 061105 (2010)
72. Pan, R.K., Kivelä, M., Saramäki, J., Kaski, K., Kertész, J.: Using explosive percolation in analysis of real-world networks. Phys. Rev. E **83**, 046112 (2011)
73. da Costa, R.A., Dorogovtsev, S.N., Goltsev, A.V., Mendes, J.F.F.: Explosive percolation transition is actually continuous. Phys. Rev. Lett. **105**, 255701 (2010)
74. da Costa, R.A., Dorogovtsev, S.N., Goltsev, A.V., Mendes, J.F.F.: Critical exponents of the explosive percolation transition. ArXiv:1402.4450
75. Grassberger, P., Christensen, C., Bizhani, G., Son, S.-W., Paczuski, M.: Explosive percolation is continuous, but with unusual finite size behavior. Phys. Rev. Lett. **106**, 225701 (2011)
76. Nagler, J., Levina, A., Timme, M.: Impact of single links in competitive percolation. Nat. Phys. **7**, 265 (2011)

77. Lee, H.K., Kim, B.J., Park, H.: Continuity of the explosive percolation transition. Phys. Rev. E **84**, 020101(R) (2011)
78. Fan, J., Liu, M., Li, L., Chen, X: Continuous percolation phase transitions of random networks under a generalized Achlioptas process. Phys. Rev. E **85**, 061110 (2012)
79. Liu, M., Fan, J., Li, L., Chen, X.: Continuous percolation phase transitions of two-dimensional lattice networks under a generalized Achlioptas Process. Eur. Phys. J. B **85**, 132 (2012)
80. Bastas, N., Kosmidis, K., Argyrakis, P.: Explosive site percolation and finite-size hysteresis. Phys. Rev. E **84**, 066112 (2011)
81. Riordan, O., Warnke, L.: Explosive percolation is continuous. Science **333**, 322–324 (2011)
82. Riordan, O., Warnke, L.: Achlioptas process phase transitions are continuous. Ann. Appl. Probab. **22**, 1450–1464 (2012)
83. Friedman, E.J., Landsberg, A.S.: Construction and analysis of random networks with explosive percolation. Phys. Rev. Lett. **103**, 255701 (2009)
84. Moreira, A.A., Oliveira, E.A., Reis, S.D.S., Herrmann, H.J., Andrade Jr, J.S.: Hamiltonian approach for explosive percolation. Phys. Rev. E **81**, 040101(R) (2010)
85. Cho, Y.S., Kahng, B.: Suppression effect on explosive percolation. Phys. Rev. Lett. **107**, 275703 (2011)
86. Bizhani, G., Paczuski, M., Grassberger, P.: Discontinuous percolation transitions in epidemic processes, surface depinning in random media, and Hamiltonian random graphs. Phys. Rev. E **86**, 011128 (2012)
87. Boettcher, S., Singh, V., Ziff, R.M.: Ordinary percolation with discontinuous transitions. Nat. Commun. **3**, 787 (2012)
88. Cao, L., Schwarz, J.M.: Correlated percolation and tricriticality. Phys. Rev. E **86**, 061131 (2012)
89. Panagiotou, K., Spöhel, R., Steger, A., Thomas, H.: Explosive percolation in Erdös-Rényi-like random graph processes. Electron. Notes Discrete Math. **38**, 699–704 (2011)
90. Araújo, N.A.M., Herrmann, H.J.: Explosive percolation via control of the largest cluster. Phys. Rev. Lett. **105**, 035701 (2010)
91. Schrenk, K.J., Araújo, N.A.M., Herrmann, H.J.: Gaussian model of explosive percolation in three and higher dimensions. Phys. Rev. E **84**, 041136 (2011)
92. Chen, W., D'Souza, R.M.: Explosive percolation with multiple giant components. Phys. Rev. Lett. **106**, 115701 (2011)
93. Buldyrev, S.V., Parshani, R., Paul, G., Stanley, H.E., Havlin, S.: Catastrophic cascade of failures in interdependent networks. Nature **464**, 1025–1028 (2010)
94. Bastas, N., Giazitzidis, P., Maragakis, M., Kosmidis, K.: Explosive percolation: unusual transitions of a simple model. Phys. A **407**, 54–65 (2014)
95. Chen, W., Zheng, Z., D'Souza, R.M.: Deriving an underlying mechanism for discontinuous percolation. Europhys. Lett. **100**, 66006 (2012)
96. Chen, W., Schröder, M., D'Souza, R. M., Sornette, D., Nagler, J.: Microtransition Cascades to Percolation. Phys. Rev. Lett. **112**, 155701 (2014)
97. Chen, W., Nagler, J., Cheng, X., Jin, X., Shen, H., Zheng, Z., D'Souza, R.M.: Phase transitions in supercritical explosive percolation. Phys. Rev. E **87**, 052130 (2013)
98. Nagler, J., Tiessen, T., Gutch, H.W.: Continuous percolation with discontinuities. Phys. Rev. X **2**, 031009 (2012)
99. Schröder, M., Ebrahimnazhad Rahbari, S.H., Nagler, J.: Crackling noise in fractional percolation. Nat. Commun. **4**, 2222 (2013)
100. Bianconi, G., Dorogovtsev, S.N.: Multiple percolation transitions in a configuration model of network of networks. ArXiv:1402.0218
101. Chen, W., Cheng, X., Chung, N.N., Zheng, Z., D'Souza, R.M., Nagler, J.: Phase transitions in supercritical explosive percolation. Phys. Rev. E **88**, 042152 (2013)

Chapter 2
Discontinuous Explosive Percolation with Multiple Giant Components

2.1 Introduction

The percolation phase transition models the onset of large-scale connectivity in lattices or networks, in systems ranging from porous media, to resistor networks, to epidemic spreading [1–4]. Percolation was considered a robust second-order transition until a variant with a choice between edges was shown to result in a seemingly discontinuous transition [5]. Subsequent studies have shown similar results for scale-free networks [6, 7], lattices [8, 9], local cluster aggregation models [10], single-edge addition models [11, 12], and models which control only the largest component [13]. It seems a fundamental requirement that in the subcritical regime the evolution mechanism produces many clusters, which are relatively large, though sublinear, in size [10, 11, 14]. Most recently, the notions of "strongly" versus "weakly" discontinuous transitions have been introduced [15], with the model studied in [5] showing weakly discontinuous characteristics,, while an idealized deterministic "most explosive" model [14, 16] is strongly discontinuous. Here, we analyze and extend a related model by Bohman, Frieze, and Wormald (BFW) [17], which predates the more recent work, and show that surprisingly [18], multiple stable giant components can coexist and that the percolation transition is strongly discontinuous.

2.2 Results

The "most explosive" deterministic process [14–16] begins with n isolated nodes, with n set to a power of two for convenience. Edges that connect pairs of isolated nodes, creating components of size $k = 2$, are added sequentially until no isolated nodes remain. The cutoff k is then doubled and edges leading to components of size $k = 4$ are added sequentially, until all components have size $k = 4$. k is then doubled yet again and the process iterated. By the end of the phase when $k = n/2$, only two components remain, each with size $n/2$. The addition of the next edge connects those

W. Chen, *Explosive Percolation in Random Networks*, Springer Theses,
DOI: 10.1007/978-3-662-43739-1_2

two components during which the size of the largest component jumps in value by $n/2$. Letting t denote the number of edges added to the graph, we define the critical t_c as the single edge whose addition produces the largest jump in size of the largest component denoted $\Delta C_{\max}$; here $t_c = (n-1)$, with $\Delta C_{\max} = n/2$.

The BFW model begins with a collection of n isolated nodes (with n any integer) and also proceeds in phases starting with $k = 2$. Edges are sampled one-at-a-time, uniformly at random from the complete graph. If an edge would lead to formation of a component of size less than or equal to k it is accepted. The edge is otherwise rejected provided that the fraction of accepted edges is greater than or equal to a function $g(k) = 1/2 + (2k)^{-1/2}$. Once the next edge rejection event would lead to the fraction of accepted nodes dropping below $g(k)$, the phase is augmented to $k+1$, with explicit details given below. Note asymptotically, $\lim_{k\to\infty} g(k) = \alpha$ with $\alpha = 1/2$.

BFW established rigorous results whereby setting $g(200) = 1/2$, all components are no larger than $k = 200$ nodes (i.e., no giant component exists) when $m = 0.96689n$ edges out of $2m$ sequentially sampled random edges have been added to graph. They further establish that a giant component must exist by the time $m = c^*n$ out of $2m$ sampled edges have been added, with $c^* \in [0.9792, 0.9793]$. Yet, they did not analyze the details of the percolation transition. We show that their model leads to the simultaneous emergence of two giant components (each of size greater than 40% of all the nodes), and show analytically the stability of the two giants throughout the subsequent graph evolution. We then generalize the BFW model by allowing the asymptotic fraction of accepted edges, α, to be a parameter and show that α determines the number of stable giant components that emerge and that, in general, smaller values of α lead to a more delayed and more explosive transition.

Stating the BFW algorithm in detail, let k denote the stage and n the number of nodes. Let u denote the total number of edges sampled, A the set of accepted edges (initialized to $A = \emptyset$), and $t = |A|$ the number of accepted edges. At each step u, an edge e_u is sampled uniformly at random from the complete graph generated by the n nodes, and the following algorithm iterated:

```
Set l = maximum size component in A ∪ {e_u}
  if (l ≤ k){
    A ← A ∪ {e_u}
    u ← u + 1 }
  else if (t/u < g(k)){ k ← k + 1 }
  else{ u ← u + 1 }
```

Thus, while $t/u < g(k)$, k is augmented repeatedly until either k becomes large enough that edge e_u is accepted or $g(k)$ decreases sufficiently that edge e_u can be rejected at which point step u ends. Note $g(k) = 1$ requires that all edges be accepted, equivalent to Erdös-Rényi [19].

We numerically implement the BFW model, and measure the fraction of nodes in both the largest and second largest component, denoted C_1 and C_2, as a function of edge density t/n. As shown in Fig. 2.1a, two giants appear at the same critical point and remain distinct. To establish that the BFW model shows a seemingly

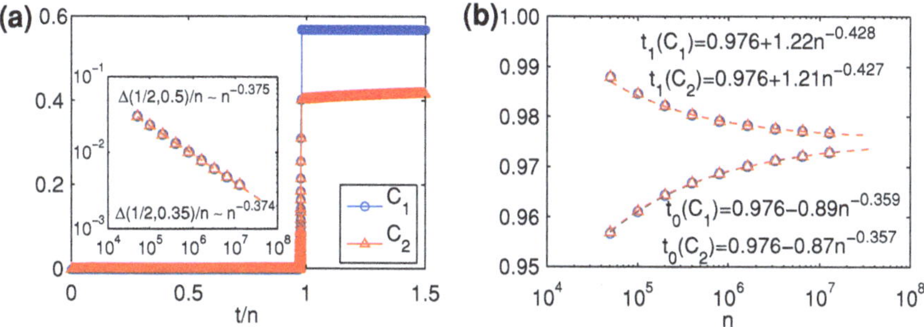

Fig. 2.1 The BFW model. **a** C_1 and C_2 versus edge density, t/n, for $n = 10^6$, showing the emergence of two stable giant components. Inset: $\Delta(1/2, 0.5)/n$, $\Delta(1/2, 0.35)/n$ for C_1 and C_2 versus n. **b** The *lower* and *upper* boundaries of Δ/n for both C_1 and C_2, where data points are averages over 200 to 2,000 realizations and dashed lines are best fits

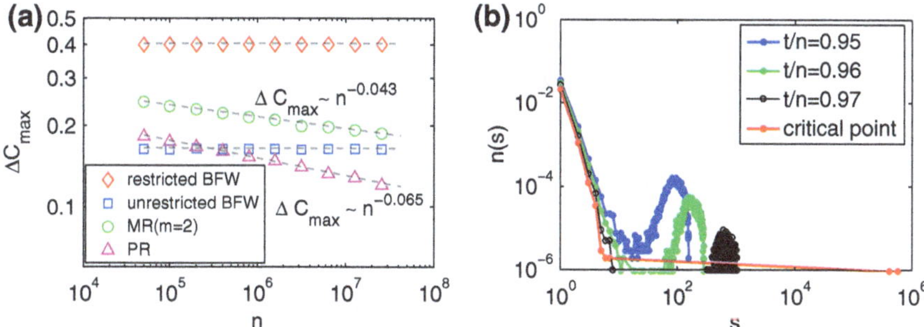

Fig. 2.2 **a** $\Delta C_{\max}$, the biggest single edge increase in C_1, is independent of system size n for BFW and for the restricted BFW process, indicating these transitions are strongly discontinuous. **b** Evolution of the distribution of $n(s)$, the fraction of components of size s, for the BFW model

discontinuous transition we apply the numerical method introduced in [5]. Let $t_0^{C_i}(n)$ denote the last accepted edge for which $C_i n \leq n^\gamma$ and $t_1^{C_i}(n)$ the first accepted edge with $C_i n \geq An$, where n is system size and γ and A are parameters. $\Delta^i(\gamma, A) = t_1^{C_i}(n) - t_0^{C_i}(n)$ denotes the number of accepted edges required between these two points. As shown inset to Fig. 2.3a, $\Delta^i(\gamma, A)/n \propto n^{-0.375}$, showing that $t_0^{C_i}(n)/n$ and $t_1^{C_i}(n)/n$ converge to same limiting value $t_c/n = 0.976$ (Fig. 2.1b) for both C_1 and C_2.

The discontinuous nature is more explicit in Fig. 2.2a showing $\Delta C_{\max}$, the largest increase of the largest component due to addition of a single edge, versus n (blue squares are BFW). $\Delta C_{\max}$ is independent of n (strongly discontinuous). Essentially, the same value of $\Delta C_{\max}$ is always observed and results from the second and third components merging together to overtake what was previously the largest. The model studied in [5] (PR) shows a decrease with n (weakly discontinuous), with scaling $n^{-0.065}$ as also recently observed in [12, 15].

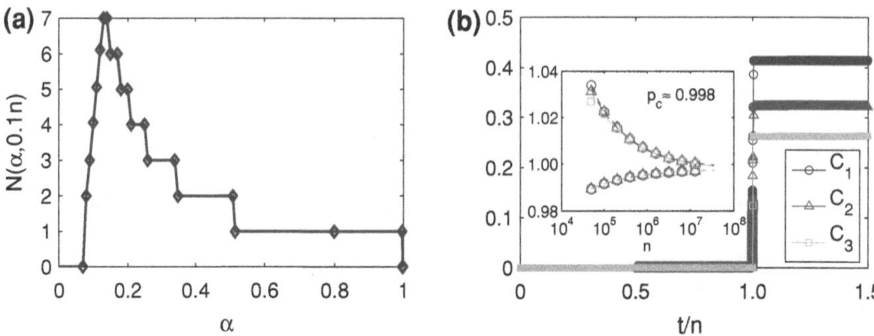

Fig. 2.3 **a** Number of stable giant components of size greater than $0.1n$ versus α for $n = 10^6$. **b** When $\alpha = 0.3$, three giant components emerge simultaneously. Inset shows convergence of the upper and lower boundaries of Δ/n for C_1, C_2, C_3

The key to coexisting multiple giant components is the high probability of sampling internal-cluster links in the super-critical region which, by definition, do not increase the component size. We formalize this by first introducing a function $P(k, t, n)$ defined as the probability of sampling a random link, which leads to a component of size no larger than stage k at step t for system size n:

$$P(k, t, n) = \sum_i C_i^2 + 2 \sum_{C_i + C_j \leq k/n} C_i C_j \tag{2.1}$$

where C_i denotes the fraction of nodes in component i. The first term on the right-hand side is the probability of randomly sampling internal-cluster links in all components. The second term is the probability of sampling spanning-cluster links, which lead to a component of size no larger than k. This is valid for any configuration in phase k. We also consider $S(k, n)$, the probability of sampling random links, which lead to components of size no larger than k if all possible spanning-cluster links are added in stage k. Thus,

$$S(k, n) = \sum_i C_i^2. \tag{2.2}$$

(Note the values of the C_i's in Eqs. (2.1) and (2.2) can differ from each other.) For any specific stage k, it is easy to show that $P(k, t, n) \geq S(k, n)$ since, if t increases, $P(k, t, n)$ can only decrease or stay the same. More explicitly, if an internal-cluster link is added then $P(k, t, n)$ is invariant, while if a spanning-cluster link is added between components i and j then the first term increases by $(C_i + C_j)^2 - (C_i^2 + C_j^2) = 2C_i C_j$ and second term decreases by at least $2C_i C_j$. (Additional decreases result if there exist components l satisfying $C_i + C_l \leq k/n$, but with $(C_i + C_j) + C_l > k/n$).

Focusing now on the critical region, let k^* denote the value of k at t_c. Numerical results for a variety of system sizes show that at t_c when $n > 10^6$, $k^*/n \sim 0.570$, $C_1 \sim 0.570$ and $C_2 \sim 0.405$ with error bars of order $\mathcal{O}(10^{-4})$ obtained over 30 to 300 realizations dependent on n. Thus, the remaining components have total size density $\sum_{i\geq 3} C_i = 0.025$. We can establish a uniform lower bound on $S(k,n)$ for all $k \geq k^*$ using the simple intuition that under the normalization condition $\sum_i C_i = 1$, $S(k,n) = \sum_i C_i^2$ is minimized when the number of components are as numerous as possible and of similar size. Given that $\sum_{i\geq 3} C_i < C_1 - C_2$, the lower bound on $S(k,n)$ is if all small components connect to C_2. Then $P(k,t,n) \geq S(k,n) \geq C_1^2 + (C_2 + \sum_{i\geq 3} C_i)^2 = 0.570^2 + (0.405 + 0.025)^2 \sim 0.510$, so for any stage $k \geq k^*(n)$, we have

$$P(k,t,n) > \alpha = 1/2. \tag{2.3}$$

So for $k \geq k^*$, the expected fraction of accepted links approaches a positive value strictly larger than α.

Having established that in expectation $P(k,t,n) > \alpha$, for $k \geq k^*$, we need to explicitly consider what happens if an edge connecting the two giant components is sampled in this regime. Here $(C_1 + C_2) > k/n \geq k^*/n$ and, by definition, $t/u \geq g(k)$. Consider the case when edge e_{u+1} connects C_1 and C_2. If $t/(u+1) \geq g(k)$ the edge is simply rejected. But if $t/(u+1) < g(k)$ either k needs to increase until the edge is accommodated, or (as we show next) a small increase in k quickly leads to $t/(u+1) \geq g(k)$. Setting t/u to the smallest value possible:

$$\frac{t}{u} = \frac{1}{2} + \sqrt{\frac{1}{2k}} \tag{2.4}$$

Differentiating both sides in Eq. (2.4) by k we find that

$$\frac{du}{d(k/n)} = \frac{1}{2\sqrt{2}(1/2 + \sqrt{1/2k})^2} \frac{nt}{k^{3/2}}. \tag{2.5}$$

After the critical point we know that $t \sim \mathcal{O}(n)$ and the stage $k \sim C_1 n \sim \mathcal{O}(n)$. Thus, from Eq. (2.5) it follows that $\frac{du}{d(k/n)} \sim \mathcal{O}(n^{1/2})$ as $n \to \infty$ implying that an $\mathcal{O}(n^{-1/2})$ increase in k/n results in $t/(u+1) > g(k)$, so the link which would lead to merging C_1 and C_2 is rejected and the two giant components are stable throughout the subsequent evolution. We verify this numerically. Letting $\bar{k}(n)$ denote the largest value of the stage ever attained for system size n, we find $(\bar{k}(n) - k^*(n))/n \sim n^{-\gamma}$ with $\gamma = 0.46 \pm 0.03$, and as $n \to \infty$, $k^*(n)/n$ and $\bar{k}(n)/n$ converge to the same limiting value of approximately 0.570.

The BFW model samples edges uniformly at random from the complete graph. If we restrict the process to sampling only edges that span distinct clusters, we observe that two components with the same $C_1 = 0.570$ and $C_2 = 0.405$ values coexist for several edge additions before merging together. When they do merge the largest jump in C_1, equal to the size of C_2, occurs. This is a strongly discontinuous transition as shown in Fig. 2.2a (the red diamonds) with gap size equal to 0.405.

We now generalize the BFW model so that $g(k) = \alpha + (2k)^{-1/2}$ (i.e., the asymptotic fraction of accepted links is now a parameter α). For the unrestricted process (sampling from the complete graph) we find that α controls the number of stable giant components. Let $N(\alpha, m)$ denote the number of stable giant components with size larger than m, which appear at the critical point and remain throughout the subsequent evolution. Figure 2.3a shows $N(\alpha, 0.1n)$ versus α for the unrestricted process, with system size 10^6 and each data point averaged over 100 independent realizations (showing no fluctuations). As α first decreases from $\alpha = 1$, $N(\alpha, 0.1n)$ increases, going from one giant to two at $\alpha = 0.511 \pm 0.003$. Then, once $\alpha < 0.11$, $N(\alpha, 0.1n)$ decreases. (Using a less stringent criteria that considers all macroscopic components $C_i n > cn$ where $c > 0$ a "giant", then $N(\alpha, cn)$ actually continues increasing.)

The same reasoning that applied to the original BFW model can be used here to show the stability of the multiple giants. Once $k \geq k^*$, in expectation $P(k, t, n) > \alpha$. Likewise, once $k \geq k^*$, $\frac{du}{d(k/n)} \sim \mathcal{O}(n^{1/2})$, so k/n increases very slowly and the process frequently samples new links and rejects links that merge any two giants. For example, if $\alpha = 0.3$, $N(\alpha, 0.1n) = 3$ with $C_1 = 0.414$, $C_2 = 0.321$, $C_3 = 0.265$, so $P(k, t, n) \geq C_1^2 + C_2^2 + C_3^2 \sim 0.345 > \alpha = 0.3$ when $k \geq k^*(n)$. See Fig. 2.3b for details.

In Fig. 2.4a we show the typical evolution for the unrestricted BFW process for various values of α in the regime where only one giant component emerges. We measure the scaling window $\Delta(\gamma, A)$, as discussed earlier, and find that smaller α leads to a more "explosive" transition in that A is larger and the scaling window shrinks more quickly. Explicitly, for $\alpha = 0.7, 0.8, 0.9$ (and setting $\gamma = 1/2$), we find respectively that $A = 0.9, 0.8, 0.7$ and $t_c \approx 0.915, 0.862, 0.780$. This delayed and more explosive nature with smaller α is intuitive in that the more links are rejected at each stage, the longer one stays in that stage, resulting in more components of size $C_i n \sim k$.

Figure 2.4b shows the analogous behavior for the restricted BFW process (where only edges that span components are considered). The delayed and more explosive nature of the transition with decreasing α is also observed here. We also note that the location of t_c is not affected. For instance, for $\alpha = 0.3, 0.5, 0.7, 0.8, 0.9$ we find

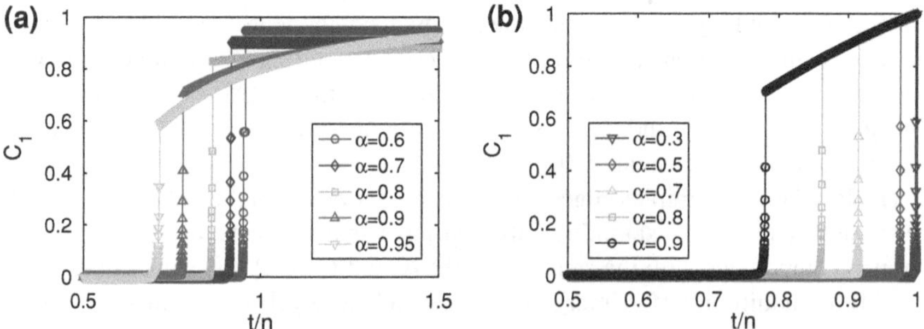

Fig. 2.4 The results of varying α on **a** the unrestricted BFW process in the regime where only one giant component emerges, and **b** the restricted BFW process. Both (*a*) and (*b*) show increasing delay and larger $\Delta C_{\max}$ for smaller α

that $t_c \approx 0.998, 0.976, 0.915, 0.862, 0.780$ for both the restricted and unrestricted processes.

The behavior of the restricted process can also be explained via Eq. (2.1). Here, because intra-cluster links are not allowed, the first term on the right-hand side vanishes. If the stage stops at some $k_0 < n$, then the second term on right side of Eq. (2.1) decreases to 0, which makes $P(k_0, t, n) < \alpha$ and stage keeps on growing until the two giants merge together. The restriction on sampled links does not change the sublinear nature of $\Delta(\gamma, A)$. In fact, we find that $\Delta(\gamma, A)$ for the general BFW model is sublinear in n for all $\alpha \in (0, 0.97]$, regardless of whether link-sampling is restricted or unrestricted, indicating the transition is discontinuous in all cases.

2.3 Summary and Discussion

In summary, we have analyzed the critical behavior of the BFW model and find that two stable giant components emerge at the same critical point in a strongly discontinuous transition (Fig. 2.2a). If we restrict the sampled links to only spanning-cluster edges, multiple giants coexist for a few moments until they merge together in a larger discontinuous jump (Fig. 2.2a) and ultimately only one giant component emerges. We further generalize BFW by making the asymptotic fraction of accepted links a parameter α, and find that number of stable giant components increases while α decreases.

The existence of multiple stable macroscopic components is surprising and unanticipated [18], and has not been previously observed in stochastic percolation. A model of cluster aggregation where largest clusters are occasionally "frozen" and prevented from growing, does lead to multiple giant components [20], however the freezing is imposed on the system. Simple algorithms that generate multiple giant components may create a new range of applications. In addition to providing insight and a potential mechanism for controlling gel sizes during polymerization [20], they may be useful for creating communication networks consisting of multiple large components operating on different frequencies or for analyzing epidemic infections simultaneously arising in distinct, independent groups. In the unrestricted process (the original BFW process) multiple links between two nodes and self-loops are allowed. It would be important to understand what happens when only links not yet added to the graph are sampled for finite networks. In the asymptotic size limit, there should be no difference as there are $O(n^2)$ available edges and $O(n)$ added edges.

The nature of the transition observed in [5] was recently analyzed using cluster aggregation models with choice, where a set of candidate edges are simultaneously inspected at each step [21]. The mechanism here, in contrast, inspects one edge individually at each time. As shown in Fig. 2.2a the models in [21] (labeled MR (m = 2)) and in [5] (labeled PR) show weakly discontinuous transitions, where $\Delta C_{\max}$ decreases with system size. In contrast both the restricted and unrestricted BFW models are strongly discontinuous, with a jump independent of system size. Finally, we show the evolution of the component size distribution for the original BFW model in Fig. 2.2b ($n(s)$ is the fraction of components of size s). This bimodal distribution has a large right-hand tail, which deviates from a power law. Whether it would show a power law at t_c as $n \to \infty$ is not obvious.

References

1. Stauffer, D., Aharony, A.: Introduction to Percolation Theory. Taylor & Francis, London (1994)
2. Sahimi, M.: Applications of Percolation Theory. Taylor & Francis, London (1994)
3. Pastor-Satorras, R., Vespignani, A.: Epidemic spreading in scale-free networks. Phys. Rev. Lett. **86**, 3200–3203 (2001)
4. Moore, C., Newman, M.E.J.: Epidemics and percolation in small-world networks. Phys. Rev. E **61**, 5678–5682 (2000)
5. Achlioptas, D., D'Souza, R.M., Spencer. J.: Explosive percolation in random networks. Science **323**, 1453–1455 (2009)
6. Cho, Y.S., Kim, J.S., Park, J.: Kahng, B, Kim, D.: Percolation transitions in scale-free networks under the Achlioptas process. Phys. Rev. Lett. **103**, 135702 (2009)
7. Radicchi, F., Fortunato, S.: Explosive percolation in scale-free networks. Phys. Rev. Lett. **103**, 168701 (2009)
8. Ziff, R.M.: Explosive growth in biased dynamic percolation on two-dimensional regular lattice networks. Phys. Rev. Lett. **103**, 045701 (2009)
9. Ziff, R.M.: Scaling behavior of explosive percolation on the square lattice. Phys. Rev. E **82**, 051105 (2010)
10. D'Souza, R.M., Mitzenmacher, M.: Local cluster aggregation models of explosive percolation. Phys. Rev. Lett. **104**, 195702 (2010)
11. Cho, Y.S., Kahng, B., Kim, D.: Cluster aggregation model for discontinuous percolation transitions. Phys. Rev. E **81**, 030103(R) (2010)
12. Manna, S.S., Chatterjee, A.: A new route to explosive percolation. Phys. A **390**, 177–182 (2011)
13. Araújo, N.A.M., Herrmann, H.J.: Explosive percolation via control of the largest cluster. Phys. Rev. Lett. **105**, 035701 (2010)
14. Friedman, E.J., Landsberg, A.S.: Construction and analysis of random networks with explosive percolation. Phys. Rev. Lett. **103**, 255701 (2009)
15. Nagler, J., Levina, A., Timme, M.: Impact of single links in competitive percolation. Nature Phys. **7**, 265–270 (2011)
16. Rozenfeld, H.D., Gallos, L.K., Makse, H.A.: Explosive percolation in the human protein homology network. Eur. Phys. J. B **75**, 305–310 (2010)
17. Bohman, T., Frieze, A., Wormald, N.C.: Avoidance of a giant component in half the edge set of a random graph. Random Struct. Algorithms **25**, 432–449 (2004)
18. Spencer, J.: The giant component: The golden anniversary. Not. AMS **57**, 720–724 (2010)
19. Erdös, P., Rényi, A.: On the evolution of random graphs. Publ. Math. Inst. Hungar. Acad. Sci. **5**, 17 (1960)
20. Ben-Naim, E., Krapivsky, P.L.: Percolation with multiple giant clusters. J. Phys. A **38**, L417–L423 (2005)
21. da Costa, R.A., Dorogovtsev, S.N., Goltsev, A.V., Mendes, J.F.F.: Explosive percolation transition is actually continuous. Phys. Rev. Lett. **105**, 255701 (2010)

Chapter 3
Deriving an Underlying Mechanism for Discontinuous Percolation Transitions

3.1 Introduction

Percolation is a theoretical underpinning for analyzing properties of networks, including epidemic thresholds, vulnerability, and robustness [1–6], with large-scale connectivity typically emerging in a smooth and continuous transition. A prototypical process begins from a collection of n isolated nodes with edges connecting pairs of nodes sequentially chosen uniformly at random and added to the graph [7]. A set of nodes connected by paths following edges is called a component, and the percolation phase transition corresponds to the first moment that there exists a component of size proportional to n (i.e., a "giant component"). A "fixed choice" variant of the simple process has gained much attention in recent years [8], where instead of a single edge, at each discrete time step a fixed number of randomly selected candidate edges are examined together, but only the edge that maximizes or minimizes a preset criteria is added to the graph. The resulting percolation transition can be extremely abrupt, with a large discontinuous jump in connectivity observed in systems with sizes larger than any real-world network (e.g., tens of billions of nodes). Yet in the $n \to \infty$ limit any fixed choice graph evolution rule leads to a continuous transition [9–12]. (which may actually be followed by discontinuous jumps arbitrarily close to the transition point [13]). Several models that lead to truly discontinuous percolation transitions are now known, for instance [14–21], yet the underlying mechanisms are not fully understood. There are many investigations underway to isolate essential ingredients that lead to a discontinuous transition, such as cooperative phenomena [22], hierarchical structures [21], correlated percolation [23], and algorithms that explicitly suppress types of growth [24].

Here, we show that a simple stochastic graph evolution process, that examines only one edge at a time, leads to a discontinuous transition and we analytically derive the simple underlying mechanism for this process: growth by overtaking. The size of the largest component does not change by direct growth, but instead it changes when two smaller components merge together and become the new largest. Overtaking is a natural growth mechanism observed in a range of systems from industrial firms [25]

W. Chen, *Explosive Percolation in Random Networks*, Springer Theses,
DOI: 10.1007/978-3-662-43739-1_3

to ecosystems [26, 27], where two smaller entities choose to cooperate (or merge) to gain competitive advantage over a previously larger entity. For the simple growth model studied here, we show that there is a control parameter (denoted β) that when small enough only allows significant growth by overtaking and leads to a discontinuous transition. But once the parameter is large enough, substantial direct growth of the giant component is allowed leading to a continuous transition. We also show that the distribution of component sizes is distinct from any previously observed. Likewise, the time evolution of the components sizes is novel. Also novel are the lack of finite-size effects for the range of β studied.

3.2 Model

The basic model we analyze was originally introduced by Bohman et al. (BFW) [28] and predates [8]. The BFW process is initialized with a collection of n isolated nodes and a cap on the maximum allowed component size set to $k = 2$. Edges are then sampled one at a time, uniformly at random from the complete graph and either added to the graph or rejected following the algorithm in Table 3.1. If an edge would lead to formation of a component of size less than or equal to k it is accepted (and we move on to sample a new edge). Otherwise, check if the fraction of accepted edges remains greater than or equal to a function $g(k) = 1/2 + (2k)^{-1/2}$. If so, the edge is simply rejected (and we move onto a new edge). If not, the cap is augmented to $k+1$ and we iterate the algorithm again. In other words, in this final case, k is augmented by one repeatedly until either k has increased sufficiently to accept the edge or $g(k)$ has decreased sufficiently that the edge can be rejected.

Here, we modify the original BFW function above such that $g(k) = 1/2 + (2k)^{-\beta}$, for $\beta \geq 0.5$, to analyze how the parameter β, which controls the rate of convergence of $g(k)$ to its asymptotic limiting value of $1/2$, affects the nature of the transition. Letting C_i denote the fraction of nodes in the ith largest component, we show both analytically and via numerical investigation that for $\beta < 1$ any significant growth in C_1 is dominated by an overtaking mechanism where smaller components merge

Table 3.1 The BFW algorithm

Set l = maximum size component in $A \cup \{e_u\}$
if $(l \leq k)$ {
$A \leftarrow A \cup \{e_u\}$
$u \leftarrow u + 1$. (Get next edge.)}
else if $(t/u \geq g(k))$ {$u \leftarrow u + 1$. (Get next edge.)}
else { $k \leftarrow k + 1$. Then repeat this block.}

At each step u, the selected edge e_u is examined via this algorithm, where u denotes the total number of edges sampled, A the set of accepted edges (initially $A = \emptyset$), and $t = |A|$ the number of accepted edges

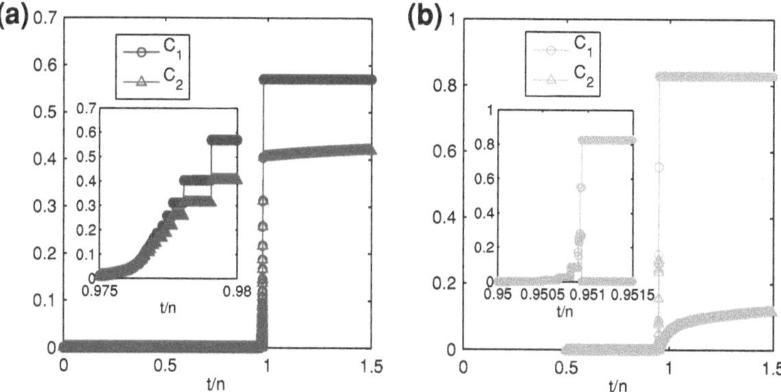

Fig. 3.1 Evolution of C_1 and C_2 versus edge density, t/n for $n = 10^6$. **a** For $\beta = 1/2$, two giant components emerge simultaneously. Inset is the behavior in the critical region showing growth via the overtaking process when what was C_1 becomes C_2. **b** A typical realization for $\beta = 2.0$. Inset shows direct growth, that C_1 and C_2 merge together, and what was C_3 becomes the new C_2

together to become the new largest component, leading to a discontinuous transition. In contrast, if $\beta > 1$ significant direct growth of C_1 is allowed and the process is dominated by stochastic fluctuations, leading to a "weakly" discontinuous transition that is likely continuous as $n \to \infty$. The typical evolution of C_1 and C_2 for $\beta = 0.5$ and $\beta = 2.0$ are shown in Fig. 3.1a, b. (The simultaneous emergence of multiple stable giant components was shown in [15], but the underlying mechanism leading to the discontinuous transition, our current focus, was not identified.)

3.3 Methods and Results

We numerically measure C_1, C_2, and C_3 throughout the evolution process for various $\beta \in [0.5, \infty]$, for a large ensemble of realizations and range of system sizes n. For each realization we define the critical point as the single edge t_c whose addition causes the largest change in the value of C_1, with this largest change denoted by ΔC^1_{max}. As shown in Fig. 3.2a, for $\beta = 0.5$, ΔC^1_{max} is independent of system size n and discontinuous. The same holds for ΔC^2_{max}, the largest jump in C_2. Yet for $\beta = 2.0$, $\Delta C^1_{\text{max}} \approx 0.285 n^{-0.0068}$. With this scaling a system of size 10^{66} would have $\Delta C^1_{\text{max}} \sim 0.1$. Following ref. [29] we label this "weakly" discontinuous, to describe the extremely slow decrease of jump size with n.

As discussed in [29], whenever a single edge is added to the evolving graph, C_1 may increase due to one of three mechanisms: (1) `direct growth`, when the largest component merges with a smaller one; (2) `doubling`, when two components both of fractional size C_1 merge (this is the largest increase possible); (3) `overtaking`, when two smaller components merge together to become the new largest. In [29] it is proven that if direct growth is strictly prohibited up to the step

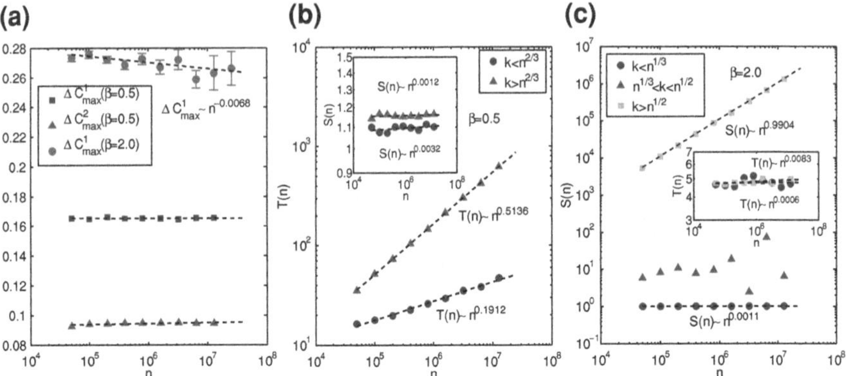

Fig. 3.2 Slow convergence leads to a strongly discontinuous transition and growth by overtaking. **a** For $\beta = 0.5$, ΔC^1_{max} and ΔC^2_{max} are independent of n and both largest components emerge in strongly discontinuous phase transitions. For $\beta = 2.0$, $\Delta C^1_{\text{max}} \sim n^{-0.0068}$, showing a weakly discontinuous phase transition. **b** For $\beta = 0.5$, main plot is $T(n)$ the number of times C_1 undergoes `direct growth` versus n, with the two regimes separated by $k = n^{1/(\beta+1)}$. (Inset) $S(n)$, the average size of component that merges with C_1 during direct growth, is essentially a constant, $S(n) \approx 1.1$ (i.e., an isolated node), in both regimes. **c** For $\beta = 2.0$, main plot is $S(n)$, showing three regimes. For $k < n^{1/(\beta+1)}$ we observe $S(n) \approx 1$. The intermediate regime is noisy. Then, once $k > n^{1/\beta}$, random edges must be accepted at times and $S(n) \sim n^{0.9904}$ (C_1 merges with other essentially macroscopic components). (Inset) $T(n)$ is essentially constant: $T(n) \approx 5$. All data points are the average over 30 to 100 independent realizations (based on system size), with error bars smaller than the symbols unless otherwise indicated

when only two components remain in the system, then a strongly discontinuous transition ensues. We next show via analytic arguments that for our modified BFW process, if $\beta < 1$ then throughout the subcritical regime direct growth only occurs when the largest component merges with an essentially isolated node and all significant growth is due to overtaking. In contrast, if $\beta > 1$ then the initial evolution is the same, but once $C_1 \sim n^{1/\beta}$, large direct growth of C_1 dominates. (Unlike [29], which requires overtaking until only two components remain, here the discontinuous transition occurs when there are still an infinite number of components in the limit of number of nodes $n \to \infty$).

Using the notation of [28], let t denote the number of accepted edges and u the total number of sampled edges. Thus, for any k (the maximum allowed component size), the fraction of accepted edges $t/u \geq g(k)$. If t/u is sufficiently large an edge leading to $C_1 n > k$ can be simply rejected. In contrast, if $t/u = g(k)$ and the next edge sampled, denoted e_{u+1}, would lead to $C_1 n > k$ that edge cannot be simply rejected since $t/(u+1) < g(k)$. One of two situations must happen, either: (i) k increases until the edge e_{u+1} is accommodated, or (ii) $g(k)$ decreases sufficiently that $t/(u+1) \geq g(k)$ and e_{u+1} is rejected. To determine which situation happens first, we need to determine the order of the smallest augmentation of k that makes $t/(u+1) > g(k)$. For $\beta \geq 0.5$ the smallest fraction of accepted edges for any k is

$$\frac{t}{u} = g(k) = \frac{1}{2} + \left(\frac{1}{2k}\right)^{\beta}. \tag{3.1}$$

Rearranging Eq. (3.1) and differentiating by k yields

$$\frac{du}{dk} = \frac{\beta}{2^{\beta}[1/2 + (1/2k)^{\beta}]^{2}} \frac{t}{k^{\beta+1}} \tag{3.2}$$

At some point in the subcritical evolution, $t \sim \mathcal{O}(n)$. (To build a component of size $\mathcal{O}(n)$ requires at least $\mathcal{O}(n)$ edges). For the BFW model with $\beta = 0.5$ it has been established rigorously that by the end of stage $k = 25$, $t/n \to 0.841$ as $n \to \infty$ [28]. We establish via numerical simulation that $t/n > 0.82$ by the end stage $k = 25$ for the full range $0.5 < \beta < 10$. Plugging $t \sim \mathcal{O}(n)$ into Eq. (3.2), once $k \sim n^{\gamma}$ (with $\gamma < 1$ for the subcritical region) then $du/dk \sim n^{1-\gamma\beta-\gamma}$. Thus, an increase in k of order $\mathcal{O}(n^{\gamma\beta+\gamma-1})$ is sufficient to ensure $t/(u+1) \geq g(k)$ and that edge e_{u+1} can be rejected. But there are different behaviors for $\beta > 1$ and $\beta < 1$.

For $\beta > 1$ there are three regimes. (i) For $\gamma \leq 1/(\beta+1)$ then $\gamma\beta + \gamma - 1 \leq 0$ so the necessary $\mathcal{O}(n^{\gamma\beta+\gamma-1})$ increase in k requires only $k \to k+1$. (ii) Then, once $1/(\beta+1) < \gamma < 1/\beta$, an increase in k of $\mathcal{O}(n^{\gamma\beta+\gamma-1}) < k \sim C_1 n$ is required. (iii) However, once $\gamma > 1/\beta$, then $\mathcal{O}(n^{\gamma\beta+\gamma-1}) > C_1 n$. Here, the required increase in k is greater than $C_1 n$, allowing C_1 to even double in size, and edge e_{u+1} must be accepted. So, once in the regime $C_1 n \sim n^{1/\beta}$ every time $t/u = g(k)$ the next edge, e_{u+1}, must be accepted. In this situation, the probability two components are merged becomes, as in Erdös-Rényi [7], proportional to the product of their sizes.

For $\beta \in [0.5, 1)$ there are only two regimes. (i) Here, again if $\gamma < 1/(\beta+1)$ then $k \to k+1$ allows edge e_{u+1} to be rejected. (ii) This regime extends until $\gamma \geq 1/(\beta+1)$, when $\gamma\beta + \gamma - 1 \geq 0$, but now we use the less strict property that $\gamma\beta + \gamma - 1 \leq \beta$ and thus $n^{\beta\gamma+\gamma-1} \leq n^{\beta} < k$. So, throughout the evolution a sublinear increase in k of at most n^{β} allows edge e_{u+1} to be rejected. The slow increase results in multiple components of size similar to C_1 throughout the evolution. In particular, once $C_1 n = \delta n$ with $\delta \ll 1$, there exist many components of size $\mathcal{O}(n)$. Order them as $C_1 n \geq C_2 n \ldots \geq C_l n$. Assuming "$>$" strictly holds (i.e., choosing only one component of each size in the case of degenerate sized components), there will be components C_l, C_{l-1} such that $C_l + C_{l-1} > C_1$. (If instead $C_l + C_{l-1} \leq C_1$, the two smaller components would merge together very quickly as the probability of randomly sampling an edge that connects them at any step u is $C_l(u)C_{l-1}(u)$, of size $\mathcal{O}(1)$, and the edge would be accepted since $C_l + C_{l-1} \leq C_1 \leq k/n$). Due to the slow increase in k, which is in increments of at most $\mathcal{O}(n^{\beta})$, there will be a point when $C_1 n < (C_l + C_{l-1})n < k < (C_l + C_{l-2})n$, allowing for growth by overtaking, when C_l and C_{l-1} merge to become the new C_1, what was C_1 becomes C_2, and what was C_{l-2} becomes C_{l-1}. The overtaking mechanism allows several large components to grow to the same order in size, which is a necessary condition to generate a strongly discontinuous percolation transition. We explicitly observe this overtaking process for $\beta = 0.5$ as shown in the inset in Fig. 3.1a.

We confirm these predictions via numerical simulations using two choices, $\beta = 1/2$ and $\beta = 2.0$. Let $S(n)$ denote the average *size* of the component C_i, which connects to C_1 via direct growth for a system of size n, and let $T(n)$ denote the number of *times* direct growth occurs. Figure 3.2b is for $\beta = 1/2$ where our analysis predicts two regimes separated by $k = n^{1/(\beta+1)} = n^{2/3}$. As shown in the inset, throughout both regimes $S(n) \approx 1.1$ is essentially a positive constant. But $T(n)$ (the main figure) shows a distinct regime change. At first $T(n) \sim n^{0.19}$. Then, in the regime starting with $k = n^{2/3}$ up to and including t_c we see a much more rapid increase, $T(n) \sim n^{0.51}$. So, we see direct growth occurring more frequently in the second regime, but the direct growth continues to be due to merging with an essentially isolated node.

Figure 3.2c is for $\beta = 2$, where our analysis predicts three regimes. Up until $k = n^{1/(\beta+1)} = n^{1/3}$ the behavior is the same as for $\beta = 0.5$ as expected since $\Delta k = 1$ is enough for an edge to be rejected and we see $S(n) \approx 1.1$. Then, in the second regime of $n^{1/3} < k < n^{1/2}$, $S(n)$ is larger and has large fluctuations. Finally, in the third regime starting from $k = n^{1/\beta} = n^{1/2}$ up until edge t_c, we see C_1 grow in large bursts, with $S(n) \sim n^{0.99}$, so C_1 merges with other essentially macroscopic components. As shown inset, $T(n)$ is essentially independent of regimes, with $T(n) \approx 5$ in the first regime and in the third, with $T(n) \leq 1$, but fluctuating in the second (not shown). The analogous three regimes and behaviors for $\beta = 3$ are shown in Fig. 3.3a. Here, once $k = n^{1/\beta} = n^{1/3}$ we see C_1 grow directly in large bursts with $S(n) \sim n^{1.01}$.

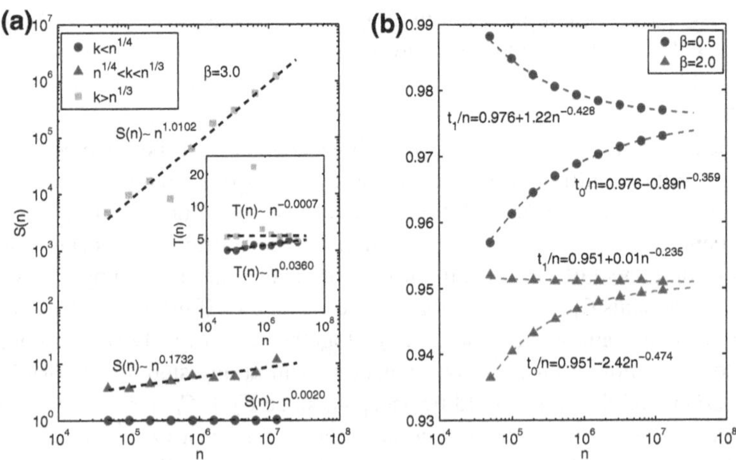

Fig. 3.3 **a** The analogous plot to Fig. 3.2c, but with $\beta = 3$: once $k > n^{1/\beta}$, the largest component merges with other macroscopic components ($S(n) \sim n^{1.0102}$). **b** Bounding the critical window from above and below to estimate t_c. For each β value the *lower line* shows the largest value of t for which $C_1 < n^{1/2}$, and the *upper line* the smallest value of t for which $C_1 > 0.5n$ for $\beta = 0.5$ and $C_1 > 0.55n$ for $\beta = 2$, yielding $t_c \approx 0.976n$ for $\beta = 0.5$ and $t_c \approx 0.951$ for $\beta = 2$

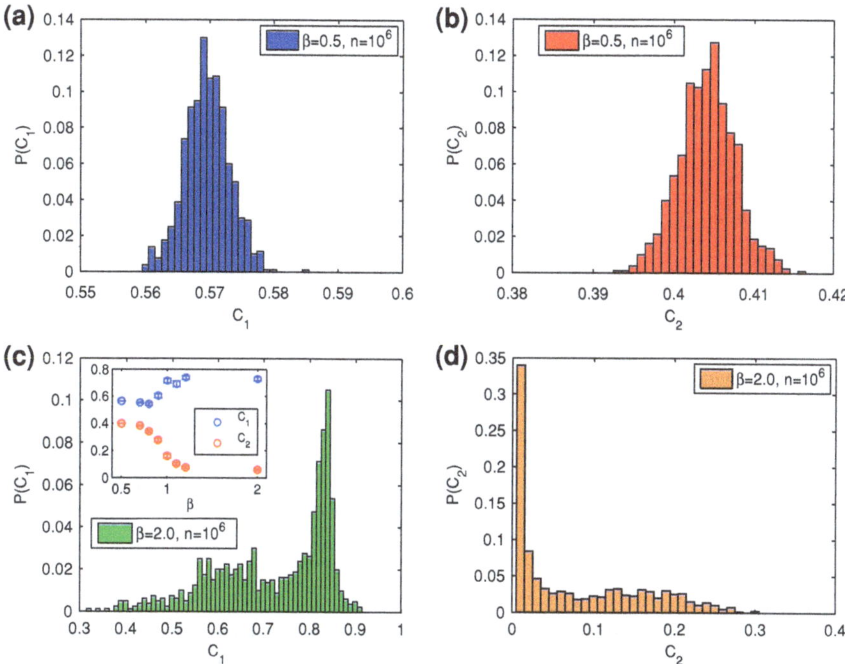

Fig. 3.4 **a–d** Distribution in values of C_1 and C_2 at t_c obtained over 100 independent realizations for $n = 10^6$. (t_c is defined as the single edge whose addition causes the biggest increase in C_1). (a) C_1 for $\beta = 0.5$. **b** C_2 for $\beta = 0.5$. **c** C_1 for $\beta = 2.0$. **d** C_2 for $\beta = 2.0$. Inset to (*c*) shows average size of C_1 and C_2 at t_c over 100 realizations for different β

For $\beta = 2.0$, due to linear increase permitted in k once $k > n^{1/2}$, and hence acceptance of random edges, we observe large fluctuations in the size of the giant components at t_c. For $\beta = 0.5$, due to the slow sublinear increase in k, the sizes of the components evolve in a predictable manner. Figure 3.4 shows these behaviors, with (a) showing C_1 and (b) C_2 observed at t_c over 100 independent realizations for $\beta = 0.5$ and (c) and (d) the equivalent for $\beta = 2$. Note for $\beta = 0.5$, $t_c \approx 0.976n$ and for $\beta = 2.0$, $t_c \approx 0.951n$, as shown in Fig. 3.3b.

A scaling analysis of the general BFW model illuminates other unique features of the model. With $\beta = 2.0$ the model exhibits critical scaling distinct from Erdös-Rényi (ER) [7] and with no finite-size effects. For $\beta = 0.5$, the model does not exhibit either critical scaling or finite-size effects. We first examine the behavior of C_1 and C_2 near t_c, as shown in Fig. 3.5a. For $\beta = 2$ we find traditional critical scaling, that $C_1, C_2 \sim (t_c - t)^{-\eta}$ with $\eta = 1.17$, the same scaling as displayed by the Product Rule (PR) [30, 31], a fixed choice edge competition rule studied in [8]. We also consider the standard finite-size scaling $C_1 = n^{-\gamma/\nu} F[(t - t_c)n^{1/\nu}]$ and perform a data collapse to determine $1/\nu = 0.49 \pm 0.02$ for BFW with $\beta = 2.0$. Note, for ER, $1/\nu = 1/3$. For BFW with $\beta = 0.5$, C_1 and C_2 show no obvious scaling behavior.

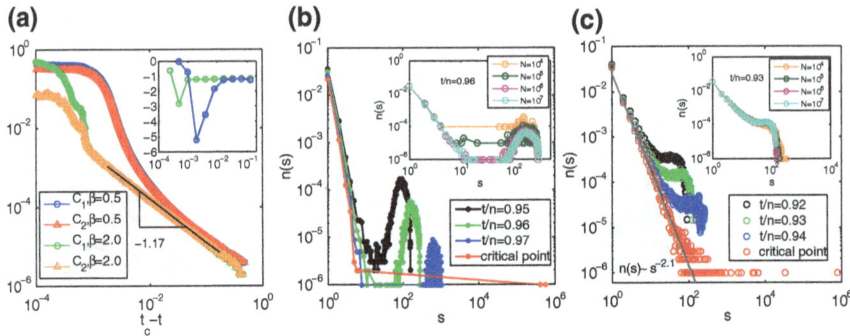

Fig. 3.5 No evident scaling behaviors for $\beta = 0.5$, whereas quantities for $\beta = 2.0$ exhibit critical scaling. **a** C_1 and C_2 versus $t_c - t$. For $\beta = 2.0$ both $C_1, C_2 \sim (t_c - t)^{-1.17}$, yet $\beta = 0.5$ shows no obvious scaling. Inset is the local slope estimate for C_1. **b** Distribution of component density $n(s)$ (number of components of size s divided by n) at different points in the evolution for $\beta = 0.5$. Inset is $n(s)$ at $t/n = 0.96$ for various n, showing no finite size effects in the location of the right hump. **c** Evolution of $n(s)$ for $\beta = 2.0$, with $n(s) \sim s^{-2.1}$ at t_c. Inset is $n(s)$ at $t/n = 0.93$ for various n, again showing no finite size effects

More importantly we study the component size density $n(s)$ (the number of components of size s divided by n). We measure the distribution of $n(s)$ at different points in the evolution up to the critical point. For $\beta = 2.0$, Fig. 3.5c, the behavior is similar to that for PR and other edge competition models with fixed choice, where at the critical point there is clear scaling behavior, $n(s) \sim s^{-\tau}$ with $\tau = 2.1$ (the same τ as for PR). Yet as shown in Fig. 3.5b, the evolution for $\beta = 0.5$ does not show any scaling. There is a pronounced right-hump, which forms early in the evolution, then moves rightward due to overtaking until there are only two large components remaining at t_c. Inset to Fig. 3.5b and c, respectively, show $n(s)$ at $t = 0.96$ and $t = 0.93$ for many different values of n. The peak of the right-hump is independent of n.

The BFW model with either $\beta = 0.5$ or $\beta = 2.0$ shows no finite-size effects, unlike ER and PR. First, for PR the location of the peak moves rightward with n, as recently shown in [12] where a finite size scaling function for PR is established. Second, Fig. 3.6a shows the fraction of edges added at the time when $C_1 n = 25$ for the first time (denoted by $\frac{t}{n}(C_1 n = 25)$) versus n for BFW, Erdös-Rényi and the PR model. For BFW, this value is independent of system size and converges to a positive constant for both $\beta = 0.5$ and 2.0, where as it decreases to 0 asymptotically for ER and PR, with $\frac{t}{n}(C_1 n = 25) \sim n^{-\tau}$, $\tau = 0.070$ and 0.015, respectively. Finally, rather than measuring t/n for fixed $C_1 n$, we can measure the value of $C_1 n$ at the time when t/n attains a specified value. Figure 3.6b shows that for BFW with both $\beta = 0.5$ and 2.0, $C_1 n$ is a positive constant for $t/n = 0.9$. Whereas for ER and PR then $C_1 n \sim n^{\theta}$ with $\theta = 0.175, 0.062$ respectively, measured in the subcritical regime for each model respectively ($t/n = 0.4$ for Erdös-Rényi, and $t/n = 0.8$ for PR.)

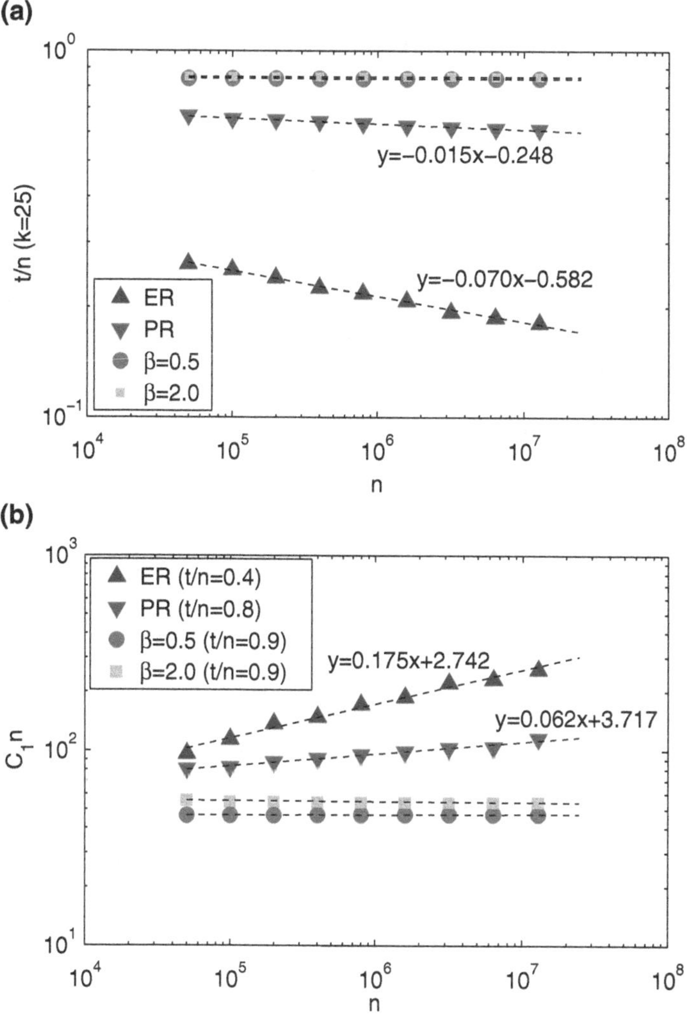

Fig. 3.6 Lack of finite size effects for BFW with $\beta = 0.5$ and $\beta = 2.0$. **a** Fraction of added edges, t/n, once $C_1 n = k = 25$, versus system size n, for ER, PR, and BFW with $\beta = 0.5$ and $\beta = 2.0$. **b** Size of largest component ($C_1 n$) versus system size n at the time when a specified fraction of edges have been added for ER, PR, and BFW with $\beta = 0.5$ and $\beta = 2.0$

3.4 Summary

In summary, we have derived the underlying mechanism that leads to the discontinuous percolation transition of the BFW model. This mechanism of growth by overtaking is a common mechanism observed in economic and ecological systems [25–27]. Algorithms such as BFW that generate multiple giant components may have a range of applications, such as controlling gel sizes during polymerization [32] or creating building blocks for modular networks. In addition, the BFW model may be simpler to implement in practice than fixed-choice edge competition models or models with cooperative interactions. We have previously shown that by varying the asymptotic fraction of accepted edges, we can control the number of resulting giant components [15]. In particular, we studied a BFW model with an acceptance function $g(k) = \alpha + (2k)^{-1/2}$ (in the discontinuous regime since $\beta = 1/2$) and showed that α controls the number of resulting giants. These giant components are stable and persist throughout the supercritical evolution: Once in the supercritical regime, there are always sufficient edges internal to components sampled that whenever an edge connecting two giant components is sampled it can be rejected. The same simple analysis holds for the most general BFW model, with $g(k) = \alpha + (2k)^{-\beta}$. From a practical perspective, we now have an algorithm for generating a specified number of stable giant components in either a discontinuous or continuous percolation transition. From a theoretical perspective, we now have an analytic understanding of the growth mechanism underlying the BFW model, which leads to a discontinuous percolation transition and to multiple stable giant components. Note, growth by overtaking is one mechanism that gives rise to discontinuous percolation, and there are also other mechanisms such as cooperation [22].

The mathematical analysis herein strongly suggests the existence of a tricritical point at $\beta = 1$, yet we cannot currently access this regime numerically. Due to the finite system size, for $\beta \in [0.7, 1.0]$, we occasionally see significant direct growth of the largest component in simulations. The rate of direct growth decreases with system size, but our current systems of size 10^7 are too small to allow a quantitative study. This is an outstanding challenge.

References

1. Stauffer, D., Aharony, A.: Introduction to Percolation Theory. Taylor & Francis, London (1994)
2. Newman, M.E.J., Watts, D.J.: Scaling and percolation in the small-world network model. Phys. Rev. E **60**, 7332–7342 (1999)
3. Moore, C., Newman, M.E.J.: Epidemics and percolation in small-world networks. Phys. Rev. E **61**, 5678–5682 (2000)
4. Cohen, R., Erez, K., Ben-Avraham, D., Havlin, S.: Resilience of the internet to random breakdowns. Phys. Rev. Lett. **85**, 4626–4628 (2000)
5. Callaway, D.S., Newman, M.E.J., Strogatz, S.H., Watts, D.J.: Network robustness and fragility: percolation on random graphs. Phys. Rev. Lett. **85**, 5468–5471 (2000)
6. Newman, M.E.J.: Networks: An Introduction. Oxford University Press, Oxford (2010)
7. Erdös, P., Rényi, A.: On the evolution of random graphs. Publ. Math. Inst. Hungar. Acad. Sci. **5**, 17 (1960)
8. Achlioptas, D., D'Souza, R.M., Spencer. J.: Explosive percolation in random networks. Science **323**, 1453–1455 (2009)
9. da Costa, R.A., Dorogovtsev, S.N., Goltsev, A.V., Mendes, J.F.F.: Explosive percolation transition is actually continuous. Phys. Rev. Lett. **105**, 255701 (2010)
10. Riordan, O., Warnke, L.: Explosive percolation is continuous. Science **333**, 322–324 (2011)

11. Grassberger, P., Christensen, C., Bizhani, G., Son, S.-W., Paczuski, M.: Explosive percolation is continuous, but with unusual finite size behavior. Phys. Rev. Lett. **106**, 225701 (2011)
12. Lee, H.K., Kim, B.J., Park, H.: Continuity of the explosive percolation transition. Phys. Rev. E **84**, 020101(R) (2011)
13. Nagler, J., Tiessen, T., Gutch, H.W.: Continuous percolation with discontinuities. Phys. Rev. X **2**, 031009 (2012)
14. Araújo, N.A.M., Herrmann, H.J.: Explosive percolation via control of the largest cluster. Phys. Rev. Lett. **105**, 035701 (2010)
15. Chen, W., D'Souza, R.M.: Explosive percolation with multiple giant components. Phys. Rev. Lett. **106**, 115701 (2011)
16. Panagiotou, K., Spöhel, R., Steger, A., Thomas, H.: Explosive percolation in Erdös-Rényi-like random graph processes. Electron. Notes Discrete Math. **38**, 699–704 (2011)
17. Araújo, N.A.M., Andrade Jr, J.S., Ziff, R.M., Herrmann, H.J.: Tricritical point in explosive percolation. Phys. Rev. Lett. **106**, 095703 (2011)
18. Cellai, D., Lawlor, A., Dawson, K.A., Gleeson, J.P.: Tricritical point in heterogeneous k-core percolation. Phys. Rev. Lett. **107**, 175703 (2011)
19. Cho, Y.S., Kahng, B.: Discontinuous percolation transitions in real physical systems. Phys. Rev. E **84**, 050102(R) (2011)
20. Schrenk, K.J., Felder, A., Deflorin, S., Araujo, N.A.M., D'Souza, R.M., Herrmann, H.J.: Bohman-Frieze-Wormald model on the lattice, yielding a discontinuous percolation transition. Phys. Rev. E **85**, 031103 (2012)
21. Boettcher, S., Singh, V., Ziff, R.M.: Ordinary percolation with discontinuous transitions. Nature Commun. **3**, 787 (2012)
22. Bizhani, G., Paczuski, M., Grassberger, P.: Discontinuous percolation transitions in epidemic processes, surface depinning in random media, and Hamiltonian random graphs. Phys. Rev. E **86**, 011128 (2012)
23. Cao, L., Schwarz, J.M.: Correlated percolation and tricriticality. Phys. Rev. E **86**, 061131 (2012)
24. Cho, Y.S., Kahng, B.: Suppression effect on explosive percolation. Phys. Rev. Lett. **107**, 275703 (2011)
25. Bengtsson, M., Kock, S.: Cooperation and competition in relationships between competitors in business networks. J. Bus. Ind. Mark. **14**, 178–194 (1999)
26. Frank, S.A.: Repression of competition and the evolution of cooperation. Evol. Int. J. Organic Evol. **57**, 693–705 (2003)
27. Hirshleifer, J.: Competition, cooperation, and conflict in economics and biology. Am. Econ. Rev. **68**, 2 (1978)
28. Bohman, T., Frieze, A., Wormald, N.C.: Avoidance of a giant component in half the edge set of a random graph. Random Struct. Algorithms **25**, 432–449 (2004)
29. Nagler, J., Levina, A., Timme, M.: Impact of single links in competitive percolation. Nature Phys. **7**, 265–270 (2011)
30. D'Souza, R.M., Mitzenmacher, M.: Local cluster aggregation models of explosive percolation. Phys. Rev. Lett. **104**, 195702 (2010)
31. Radicchi, F., Fortunato, S.: Explosive percolation: a numerical analysis. Phys. Rev. E **81**, 036110 (2010)
32. Ben-Naim, E., Krapivsky, P.L.: Percolation with multiple giant clusters. J. Phys. A **38**, L417–L423 (2005)

Chapter 4
Continuous Phase Transitions in Supercritical Explosive Percolation

4.1 Introduction

Percolation in networks, a phase transition from small, scattered components to large-scale connectivity, is heavily studied and widely applied in technological and social systems [1–4], biological networks [5, 6], epidemiology [7–10] and even dynamical models of economic systems [11, 12]. One of the most classic and widely studied models is the Erdös-Rényi random graph (ER) started from a collection of n initially isolated nodes, where edges sampled uniformly from the complete graph are sequentially added between nodes. Percolation under ER-like processes is considered a robust continuous phase transition with a unique giant component emerging at the percolation threshold [13]. Thus, altering the location and nature of the percolation phase transition has been a long-standing challenge. Three years ago, Achlioptas et al. proposed a modified ER model in which two random edges are sampled simultaneously, but only the edge connecting two components with smaller product of their sizes is added, while the other edge is discarded. The authors showed that the percolation transition can thus be delayed considerably, but when the transition eventually happens it is extremely abrupt, calling the phenomena "explosive percolation" [14]. They furthermore presented strong numerical evidence suggesting that the resulting percolation transition was, in fact, discontinuous. Many efforts were made to study similar processes on different topologies, showing that explosive percolation is observed in scale-free networks and lattices [15–19]. Although such explosive percolation phenomena resulting from the choice between a *fixed* number of random edges, as studied in [14], was later demonstrated to be actually *continuous* [20–24], several alternative models are now known to exhibit truly discontinuous percolation transitions [25–37].

Chen and D'Souza recently established that the percolation transition is discontinuous for a stochastic graph evolution model introduced by Bohman, Frieze and Wormald (the BFW model) [38], which examines a single edge at a time (rather than a model like that in [14], which examines multiple edges at each time step). In [39] they showed that the BFW model exhibits a discontinuous percolation transition leading

W. Chen, *Explosive Percolation in Random Networks*, Springer Theses,
DOI: 10.1007/978-3-662-43739-1_4

to the simultaneous emergence of multiple giant components, and that the number of stable giant components can be tuned via a parameter denoted α. In [40] the underlying mechanism leading to the discontinuous phase transition was derived, namely that growth of the largest component is dominated by overtaking when two smaller components merge together to become the new largest component. Such growth by overtaking is a natural growth mechanism observed in many complex systems from ecologies to the business world [41–43]. Additionally, the discovery of multiple giant component is unanticipated [44], and may have applications in polymerization [45], network discovery [46], and epidemiology [47].

The nature of explosive percolation phase transitions has been the topic of intense debate for the past few years [20–24, 48], but more recently, we are learning that the evolution of such processes in the supercritical regime can also be unique and of great interest. For instance, in the supercritical regime, the size of the largest component can be a nonself averaging quantity [49], furthermore, there can be multiple discontinuous jumps in the size of the largest component [50]. Here, we investigate, both numerically and analytically, the modified BFW model in the supercritical regime.

Specifically, we study the initial percolation transition and supercritical evolution of the BFW model for the value of the tunable parameter in the range $\alpha \in [0.6, 0.95]$. (This regime is simplest, as only one giant component initially emerges in a discontinuous phase transition). We show that the largest component eventually stops growing leading to the emergence of a second giant component at some delayed time, and that the second component is stable persisting throughout the subsequent evolution. Via an extensive finite size scaling analysis we establish that the second transition is continuous, but that only for $\alpha \in [0.65, 0.7]$ is this transition in the same universality class as the classic model of Erdös-Rényi. The main result we derive is that there exists a critical value α_c such that, for $\alpha \in [0.6, \alpha_c)$, $\mathbf{C_1}$ stops growing at the discontinuous percolation transition, yet for $\alpha \in (\alpha_c, 0.95]$, $\mathbf{C_1}$ continues growing for some time into the supercritical regime. Furthermore, for $\alpha < \alpha_c$, the asymptotic size of $\mathbf{C_1}$ is smaller with increasing α, leading to a smaller gap between the two percolation transitions, while for $\alpha > \alpha_c$, in contrast, the value of $\mathbf{C_1}$ is larger with increasing α, leading to a bigger gap between the two percolation transitions. As we show, α_c marks the minimal delay possible between the emergence of $\mathbf{C_1}$ and $\mathbf{C_2}$ (i.e., the smallest edge density for which $\mathbf{C_2}$ can exist).

The rest of the chapter is organized as follows. In Sect. 4.2, we discuss the BFW model and show further evidence for the discontinuous transition of the largest component. In Sect. 4.3, we analyze the growth cessation of the largest component in the supercritical regime. Section 4.4 contains an analysis of the critical behavior of the second phase transition through finite size scaling. We conclude with some discussion in Sect. 4.5.

4.2 The BFW Model and the Discontinuous Transition of the Largest Component

In this section, we illustrate the BFW model. The system is initialized with N isolated nodes and a cap, k, on the maximum allowed component size set to $k = 2$ and increased in stages as follows. Edges are sampled one-at-a-time, uniformly at random from the complete graph. If an edge would lead to formation of a component of size less than or equal to k it is accepted. Otherwise, the edge is rejected provided that the fraction of accepted edges remains greater than or equal to a function $g(k) = \alpha + (2k)^{-1/2}$, where α is a tunable parameter. Finally, if the fraction of accepted nodes would drop below $g(k)$, the cap is augmented to $k+1$, and the impact of adding the edge reevaluated against the new values, $k + 1$ and $g(k + 1)$. This final step of augmenting the cap and reevaluating the impact is iterated until either k increases sufficiently to accept the edge or $g(k)$ decreases sufficiently that the edge can be rejected. Asymptotically, the fraction of accepted edges is $\lim_{k\to\infty} g(k) = \alpha$. (Note in the original BFW model they only studied the case $\alpha = 1/2$).

Stating the BFW algorithm in detail requires some notation. Let k denote the cap size, N denote the number of nodes, u denote the total number of edges sampled, A denote the set of accepted edges, and $t = |A|$ denote the number of accepted edges. At each step u, an edge e_u is sampled uniformly at random from the complete graph generated by the N nodes and evaluated by the following algorithm, stopping once a specified number of edges, ωN, have been accepted:

Repeat until $u = \omega N$
{
  Set $l =$ **size of largest component in** $A \cup \{e_u\}$
  if $(l \leq k)$ {
    $A \leftarrow A \cup \{e_u\}$
    $u \leftarrow u + 1$ }
  else if $(t/u < g(k))$ { $k \leftarrow k + 1$ }
  else { $u \leftarrow u + 1$ }
}

It has been shown in Ref. [39] that the percolation transition of the order parameter $|\mathbf{C_1}| := C_1$, defined as the fraction of nodes in the largest component, is discontinuous in the thermodynamic limit for $\alpha \in (0, 0.97]$ (see, e.g., Fig. 4.1a). This holds regardless of whether we sample uniformly at random all edges from the complete graph or sample only edges not yet existing in the graph [39]. (The former allows intra-component edges including multiple edges and self-edges). Here, we provide additional numerical evidence for the discontinuous transition by examining three quantities: the largest jump in C_1 resulting from adding one edge, denoted as $\Delta C_{\max}$, order parameter C_1 immediately after the largest jump, denoted as P_∞, and the maximum in the second moment of the relative-size distribution of components, excluding the contribution of the largest component,

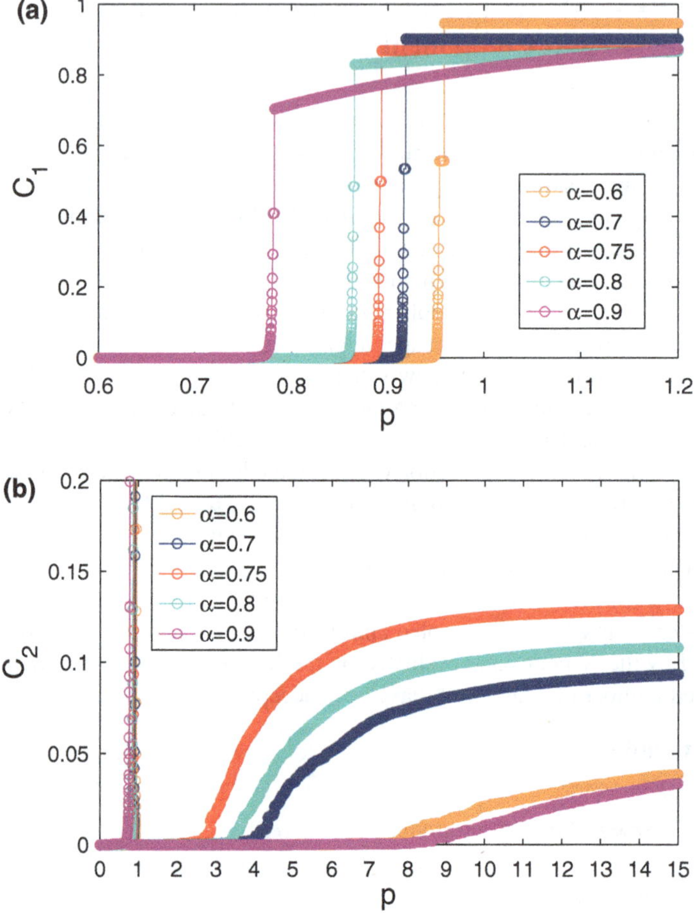

Fig. 4.1 Typical evolution of the BFW model. **a** Fraction of nodes in the largest component, the size of $\mathbf{C}_1$ as a function of edge density $p \equiv t/N$, for different values of α, showing a discontinuous phase transition at a location dependent on the value of α. As derived herein, for $\alpha > \alpha_c$ the value of $\mathbf{C}_1$ increases in the supercritical regime. **b** Fraction of nodes in the second largest component, the size of $\mathbf{C}_2$, as a function of p for different values of α, showing an instantaneous peak when $\mathbf{C}_1$ emerges followed later by a continuous phase transition. The extent of delay between the discontinuous emergence of $\mathbf{C}_1$ and the continuous emergence of $\mathbf{C}_2$ depends in the value of α The system size is $N = 10^6$

$$M_2' = M_2 - C_1^2 \tag{4.1}$$

where $M_2 = \sum_i C_i^2$ and C_i is the relative size of component i. If $\Delta C_{\max}$, P_∞, M_2' converge to some positive constant in the thermodynamic limit, the transition is discontinuous, while they vanish for continuous phase transitions [21, 26, 28, 37]. We observe $\Delta C_{\max}$, P_∞ and M_2' are all asymptotically independent of system sizes within error bars, which are shown in Fig. 4.2a–c, respectively. For large size systems

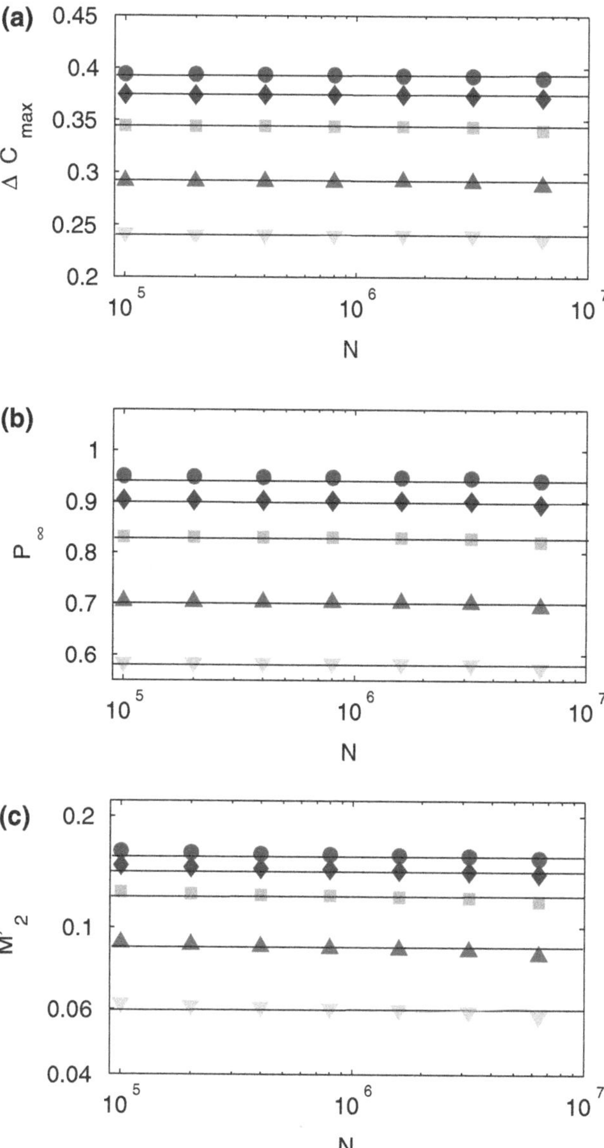

Fig. 4.2 **a** The largest jump in the size of $\mathbf{C}_1$ resulting from one edge as a function of system sizes for $\alpha = 0.6, 0.7, 0.8, 0.9, 0.95$ ordered from *top* to *bottom*. **b** The order parameter P_∞, which is C_1 immediately after the largest jump of C_1, as a function of system sizes for $\alpha = 0.6, 0.7, 0.8, 0.9, 0.95$ from *top* to *bottom*. **c** The maximum of the second moment of the cluster-size distribution M_2' as a function of system size for $\alpha = 0.6, 0.7, 0.8, 0.9, 0.95$ from *top* to *bottom*. All the quantities in (**a**), (**b**) and (**c**) are asymptotically independent of system sizes and the *solid lines* are the mean values. These results indicate that the percolation transitions are discontinuous. Each data is averaged over 500 realizations

Table 4.1 Summary of numerical results showing discontinuous phase transitions in the order parameter C_1

α	ΔC_{max}	P_∞	M_2'
0.6	0.391 ± 0.005	0.947 ± 0.004	0.154 ± 0.005
0.7	0.374 ± 0.006	0.901 ± 0.005	0.139 ± 0.006
0.8	0.344 ± 0.005	0.829 ± 0.004	0.118 ± 0.006
0.9	0.291 ± 0.005	0.702 ± 0.005	0.084 ± 0.006
0.95	0.240 ± 0.006	0.582 ± 0.005	0.057 ± 0.008

and differing values of α, we obtain the asymptotic values of ΔC_{max}, P_∞ and M_2' for $\alpha = 0.6, 0.7, 0.8, 0.9, 0.95$, respectively (Table 4.1). These numerical results indicate that the percolation transition is discontinuous for the range studied here, $\alpha \in [0.6, 0.95]$.

To estimate the percolation threshold, we implement the numerical methods proposed in Refs. [28, 37]. We measure the edge density $p_{c,1}(N)$ at which the largest jump in the size of $\mathbf{C_1}$ occurs and the edge density $p_{c,2}(N)$ at which M_2' attains maximum for different system sizes N. Extrapolating these estimators in thermodynamic limit, we obtain $p_c = 0.948 \pm 0.007, 0.915 \pm 0.008, 0.862 \pm 0.007, 0.780 \pm 0.007, 0.711 \pm 0.006$ for $\alpha = 0.6, 0.7, 0.8, 0.9, 0.95$ respectively. These results indicate the percolation threshold is larger for smaller α. We further observe asymptotic power law relation between $|p_{c,i} - p_c|(i = 1, 2)$ and system size N. In particular, $|p_{c,i} - p_c| \sim N^{-\delta}$ with $\delta = 0.3 \pm 0.1, 0.4 \pm 0.1, 0.4 \pm 0.1, 0.5 \pm 0.1, 0.5 \pm 0.1$ for $\alpha = 0.6, 0.7, 0.8, 0.9, 0.95$ respectively. The exponent $\delta = 0.5 \pm 0.1$ for $\alpha = 0.9, 0.95$ is the same as what is observed for BFW model ($\alpha = 0.5$) in square lattice within error bars [37].

4.3 Growth Cessation of the Largest Component

The discontinuous transition of the order parameter C_1 has been substantiated [39] and the underlying mechanism accounting for the discontinuous transition has been identified [40]. However, the behavior of C_1 after the largest jump (or in supercritical regime) has not been studied previously. Here, we study C_1 for values of the edge density $p \in [0, 10]$ for different system sizes. Interestingly, we observe that C_1 stops increasing either immediately after the largest jump of C_1 or at some delayed point according to specific $\alpha \in [0.6, 0.95]$. This growth cessation of the largest component in the supercritical regime has however, not been reported in other models in which any edge from the complete graph is allowed to be sampled.

Using the notation in Sect. 4.2, t denotes the number of accepted edges, and thus t/N is the edge density (also denoted throughout by the shorthand $p \equiv t/N$). Let us denote t_c as the point where the largest jump of C_1 occurs and k_c as the value of stage k at t_c. Also denote k_{final} as the final value when k stops increasing and $t(k_{final})$ as the point where k stops increasing. We measure $k(t_c)$, k_{final}, t_c/N, $t(k_{final})/N$ for

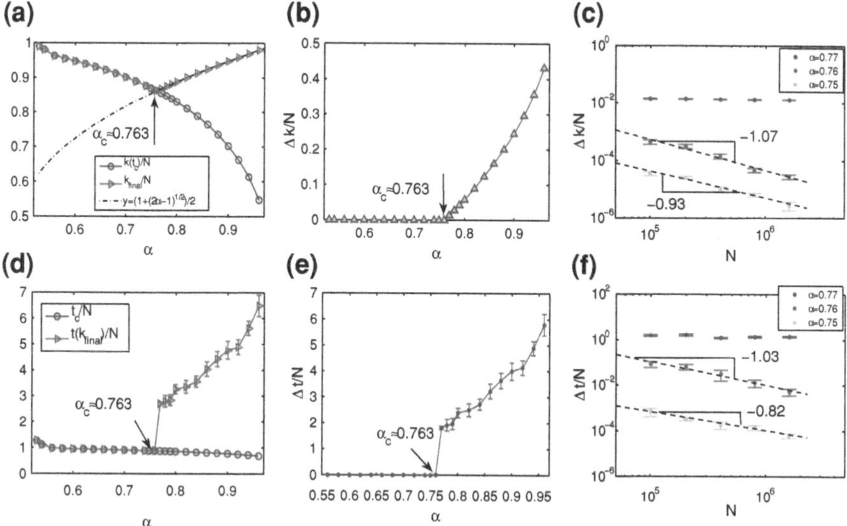

Fig. 4.3 System size is $N = 10^6$ for (**a**), (**b**), (**d**), (**e**). **a** Behavior of the cap k versus α specifically measuring $k(t_c)/N$ (blue circles) the value at the discontinuous emergence of $\mathbf{C}_1$, and k_{final}/N (*red triangles*) the value when k stops increasing. Both lines intersect the function derived in Eq. (4.4), $y = (1 + (2\alpha - 1)^{1/2})/2$, at the same point $\alpha_c \approx 0.763$. **b** $\Delta k/N = (k_{\text{final}} - k(t_c))/N$ as a function of α. **c** $\Delta k/N$ versus system size for $\alpha = 0.75, 0.76, 0.77$, showing $\Delta k/N$ converges to 0 for $\alpha = 0.75, 0.76$, while it converges to a positive constant for $\alpha = 0.77$. Further analysis shows $\alpha = 0.763 \pm 0.002$ is the transition point between the behaviors. **d** t_c/N and $t(k_{\text{final}})/N$ as a function of α. **e** Corresponding density of edges added between t_c and $t(k_{\text{final}})$, denoted as $\Delta t/N = (t(k_{\text{final}}) - t_c)/N$, is plotted as a function of α. **f** $\Delta t/N$ versus system size for $\alpha = 0.75, 0.76, 0.77$, showing $\Delta t/N$ converges to 0 asymptotically for $\alpha = 0.75, 0.76$, while it asymptotically converges to some positive constant for $\alpha = 0.77$. More detailed analysis shows $\alpha = 0.763 \pm 0.002$ is the transition point between the two behaviors

different α in a system of size $N = 10^6$. Figure 4.3a shows when $\alpha < \alpha_c \approx 0.763$, $k(t_c)$ and k_{final} overlap and both decrease with α. When $\alpha > \alpha_c$ however, $k(t_c)$ keeps on decreasing while k_{final} increases with α. Therefore, k_{final}/N reaches a minimum at $\alpha = \alpha_c$. In Fig. 4.3b, $\Delta k/N = (k_{\text{final}} - k(t_c))/N$ is plotted as a function of α, showing $\Delta k/N$ becomes finite once $\alpha > \alpha_c \approx 0.763$. Figure 4.3c shows $\Delta k/N \sim N^{-1.07}, N^{-0.93}$ for $\alpha = 0.76, 0.75$ respectively, indicating $\Delta k/n \to 0$ as $n \to \infty$. For $\alpha = 0.77$ however, $\Delta k/N$ converges to some positive constant. More precise investigation detailed below shows the transition point $\alpha_c = 0.763 \pm 0.002$. Figure 4.3d shows t_c/N and $t(k_{\text{final}})/N$ overlap when $\alpha < \alpha_c \approx 0.763$ and there is a finite gap between them once $\alpha > \alpha_c$. We observe that $t(k_{\text{final}})$ reaches a minimum at α_c. In Fig. 4.3e, $\Delta t/N = (t(k_{\text{final}}) - t_c)/N$ is plotted as a function of α, showing a sudden jump from 0 at $\alpha \approx 0.763$. In Fig. 4.3f $\Delta t/N$ is plotted as a function of N for $\alpha = 0.75, 0.76, 0.77$ where $\Delta t/N \sim N^{-1.03}, N^{-0.82}$ for $\alpha = 0.76, 0.75$ respectively but asymptotically converges to some positive constant for $\alpha = 0.77$. This numerical result shows the same transition point at $\alpha_c \approx 0.763$.

To understand the underlying mechanism accounting for the fact that the stage k and the size of the largest component stops increasing in the supercritical regime, it is useful to measure $P(k, t, N)$, defined as the probability of sampling a random edge that keeps $C_1 N \leq k$ if it is added at step t in a system of size N. If $\lim_{N\to\infty} P(k, t, N) \geq \lim_{k\to\infty} g(k) = \alpha$ always holds for $t > t_c$ where $k \sim \mathcal{O}(N)$, the fraction of accepted edges over total sampled edges t/u quickly converges to some value larger than α, so we can always discard an edge that would increase C_1, thus stage stops increasing. However, if $\lim_{N\to\infty} P(k, t, N) < \alpha$, the stage k keeps on increasing until the following condition holds at $t = t(k_{\text{final}})$

$$\lim_{N\to\infty} P(k, t, N) = \alpha. \tag{4.2}$$

We measure $k(t_c)/N$ and k_{final}/N with system sizes $N = 10^6$ for over 100 realizations. We find that at the point where the largest jump of C_1 occurs, namely t_c/N, for all $\alpha \in [0.6, 0.95]$, $k(t_c)/N = C_1 > 0.5$ averaged over 1000 realizations. Due to this fact, once $t > t_c$, the growth of $\mathbf{C_1}$ can only occur due to the mechanism of direct growth [21], in particular, the edge connecting the largest component and another smaller component results in the growth of $\mathbf{C_1}$. Therefore, the probability of sampling a random edge which leads to the growth of $\mathbf{C_1}$ is $2(1 - C_1)C_1$, and thus, $P(k, t, N)$ satisfies

$$\lim_{N\to\infty} P(k, t, N) = 1 - 2(1 - C_1)C_1. \tag{4.3}$$

Combining Eqs. (4.2) and (4.3), we obtain C_1 at $t(k_{\text{final}})$ for infinite system

$$C_1 = \frac{1 + \sqrt{2\alpha - 1}}{2}. \tag{4.4}$$

Figure 4.3a shows the dashed line of Eq. (4.4) intersects $k(t_c)/N$ and k_{final}/N at the same point $\alpha_c = 0.763$ within error bars for system size $N = 10^6$. For $\alpha < \alpha_c$, $k(t_c)/N > \frac{1+\sqrt{2\alpha-1}}{2}$ (blue circles lie above the dashed line), and $P(k, t_c, N) > \alpha$. Therefore, the stage k stops increasing immediately after the largest jump of C_1 and k_{final}/N is almost the same as $k(t_c)/N$ (blue circles and red triangles overlap). However, for $\alpha > \alpha_c$, $k(t_c)/N < \frac{1+\sqrt{2\alpha-1}}{2}$(blue circles lie below the dashed line). Therefore, both the stage k and $\mathbf{C_1}$ increase until Eq. (4.4) holds (red triangles and the dashed line overlap).

4.4 Continuous Transition of the Second Largest Component

The size of the largest component stops increasing at the point $t(k_{\text{final}})/N$ (which is either at the discontinuous phase transition point or in the supercritical regime). Yet, the second largest component continues growing at this point. Two natural and interesting questions follow. Is there a second giant component emerging at some step $t > t(k_{\text{final}})$? If such percolation transition occurs, is it a discontinuous phase transition as we find for the order parameter C_1? To answer these questions, we first numerically measure C_2 as a function of the edge density $p \equiv t/N$ for $\alpha = 0.6, 0.65, 0.7, 0.75, 0.8, 0.85, 0.9$. As shown in Fig. 4.1b, we see C_2 spike instantaneously at the first discontinuous transition, then it disappears and we observe a seemingly continuous percolation transition when C_2 again emerges at some point in the supercritical regime.

To investigate the continuous nature of percolation transition of C_2, we make use of finite size scaling [51]. If a phase transition is continuous, every variable X is believed to be scale-independent and obeys the following finite size scaling form near the percolation threshold p_c

$$X = N^{-\theta/\nu} F[(p - p_c)N^{1/\nu}] \tag{4.5}$$

where θ and ν are critical exponents, N is the system size and F is a universal function. As a result, X follows a power law at $p = p_c$, $X \sim N^{-\theta/\nu}$.

Here, we consider two variables in the percolation process: the relative size of the second largest component C_2 and the susceptibility χ_∞. The susceptibility χ_∞ is defined as the standard deviation of C_2

$$\chi_\infty = \sqrt{\langle {C_2}^2 \rangle - \langle C_2 \rangle^2}. \tag{4.6}$$

We assume that C_2 and χ_∞ obey the following scaling relations

$$C_2 = N^{-\beta/\nu} F^{(1)}[(p - p_c')N^{1/\nu}] \tag{4.7}$$

$$\chi_\infty = N^{-\gamma/\nu} F^{(2)}[(p - p_c')N^{1/\nu}] \tag{4.8}$$

where $F^{(1)}$ and $F^{(2)}$ are different universal functions; however, they are related to each other, p_c' denotes the percolation threshold of the second phase transition, and β, γ, ν are critical exponents characterizing the transition. We can deduce from the definition of χ_∞ (Eq. (4.6)) and the scaling relations in Eqs. (4.7) and (4.8) at $p = p_c'$

$$\beta/\nu = \gamma/\nu. \tag{4.9}$$

Since Eqs. (4.7) and (4.8) imply that the curves of $N^{\beta/\nu} C_2$ versus p and $N^{\gamma/\nu}\chi_\infty$ vs p for different systems sizes cross at a single point at $p = p_c'$, both the relative size of the second largest component $\mathbf{C_2}$ and the susceptibility χ_∞ can be used for the determination of percolation threshold p_c'. More precisely, we implement the minimization technique in Ref. [52]. For each system size N, we use Monte Carlo data for C_2 versus p and χ_∞ versus p to create functions $g_N = N^{\beta/\nu} C_2$ and $l_N = N^{\gamma/\nu}\chi_\infty$ respectively. The value of p that minimizes the objective function $P = \sum_{Ni<Nj}(g_{Ni} - g_{Nj})^2$ is the estimation of percolation threshold p_c'. Similarly,

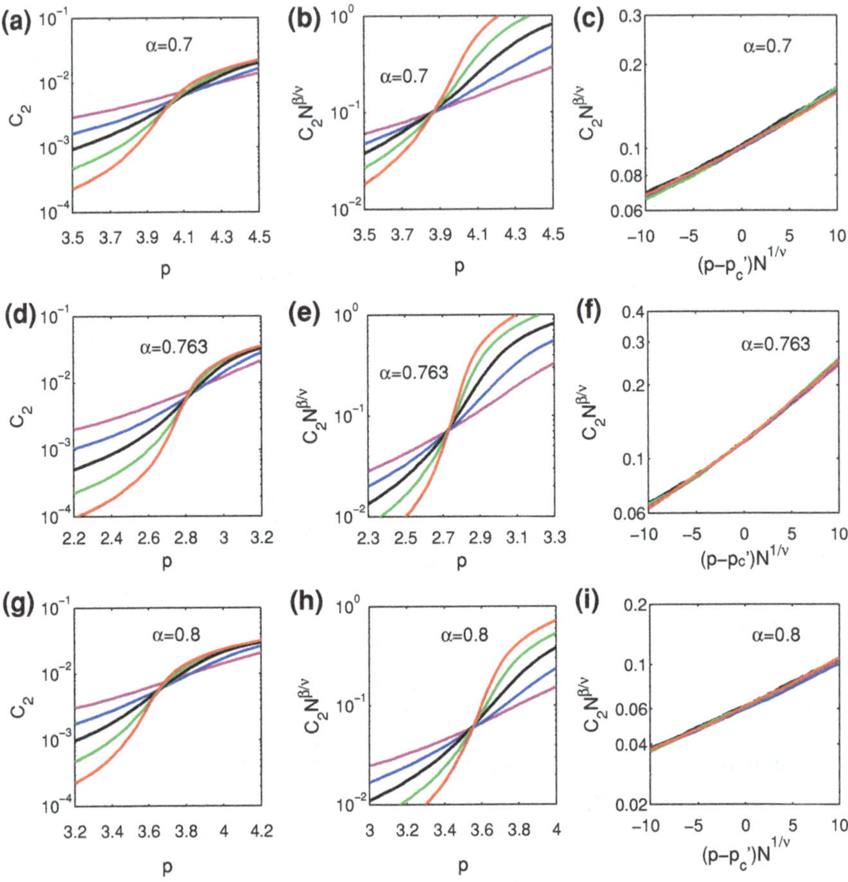

Fig. 4.4 In **a**, **d**, **g**, fraction of nodes in the second largest component, C_2 is plotted as a function of edge density p for $\alpha = 0.7, 0.763, 0.8$, respectively. **b**, **e**, **h** show their rescaling $C_2 N^{\beta/\nu}$, respectively. **c**, **f**, **i** show $C_2 N^{\beta/\nu}$ versus $(p - p_c')N^{1/\nu}$, indicating the validity of Eq. (4.7). The system size goes from $N = 20000$ (*magenta*) to 1620000 (*red*) via successive doubling

the value of p, which minimizes the objective function $Q = \sum_{Ni<Nj}(l_{Ni} - l_{Nj})^2$, is an alternative way to estimate p_c'. In our following numerical simulations, we find a perfect agreement within error bars between the two different methods.

In Fig. 4.4a, d, and g, we plot the order parameter C_2 as a function of edge density p, for $\alpha = 0.7, 0.763, 0.8$ in BFW model, respectively. Five curves in each figure stand for Monte Carlo data averaged over 1000 realizations for five different system sizes, namely $N = 20000, 60000, 180000, 540000, 1620000$. To identify the percolation threshold p_c' of C_2 and two scaling exponents β/ν, γ/ν, we first implement the minimization technique in the objective function P. Figure 4.4b, e, and h show that all curves of $C_2 N^{\beta/\nu}$ versus p cross at a single percolation transition point $p_c' = 3.865 \pm 0.005, 3.555 \pm 0.005, 2.725 \pm 0.003$ for $\alpha = 0.7, 0.763, 0.8$, respectively.

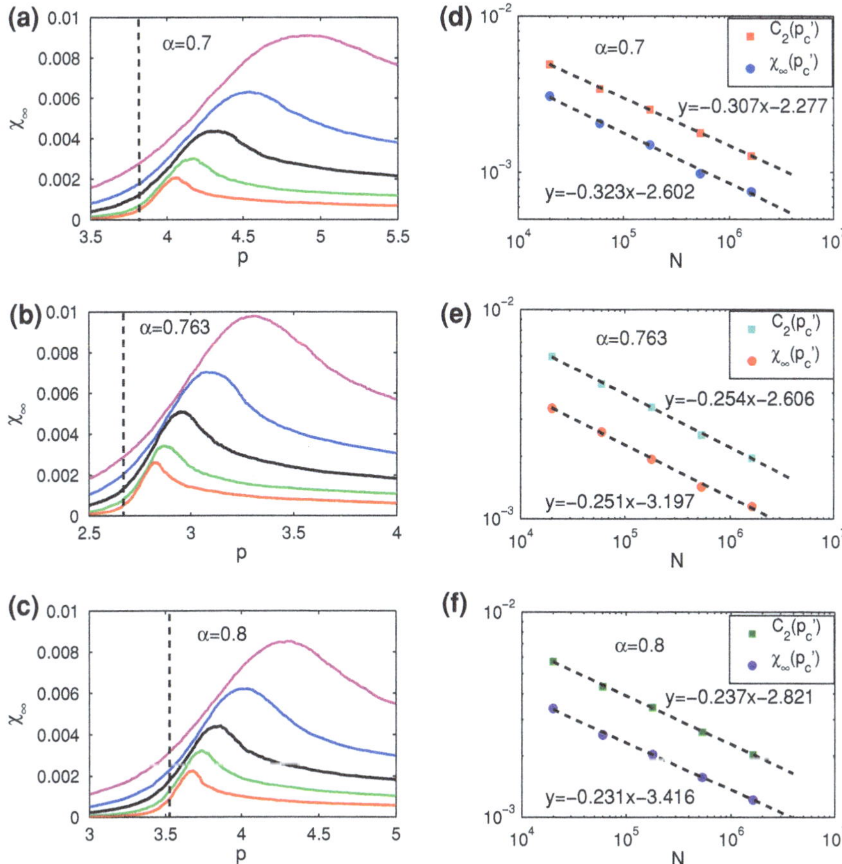

Fig. 4.5 In **a**, **b**, **c**, the susceptibility χ_∞ is plotted as a function of p for $\alpha = 0.7, 0.763, 0.8$, respectively, showing the peak of the susceptibility χ_∞ moves leftward as system size increases. The *dashed lines* are the percolation threshold obtained from numerical simulation through Eq. (4.8). The peaks of the susceptibility is moving leftward to the percolation threshold p_c' which indicates the continuous nature of the second phase transition. In **d**, **e**, **f**, C_2 and χ_∞ at the percolation threshold p_c' is plotted as a function of system sizes for $\alpha = 0.7, 0.763, 0.8$, respectively. The system size goes from $N = 20000$ (*magenta*) to 1620000 (*red*) via successive doubling

The corresponding exponent ratios β/ν are $0.31 \pm 0.03, 0.25 \pm 0.04, 0.24 \pm 0.04$ for $\alpha = 0.7, 0.763, 0.8$, respectively. In Fig. 4.4c, f, and i, the data collapses show the profiles of the universal scaling function $F^{(1)}$ of Eq. (4.7), where we find the exponent $1/\nu = 0.36 \pm 0.01, 0.36 \pm 0.01, 0.37 \pm 0.01$ for $\alpha = 0.7, 0.763, 0.8$, respectively.

In Fig. 4.5a–c, χ_∞ is plotted as a function of p, where the dashed lines show the location of percolation threshold through minimization technique in the objective function Q. We find that percolation threshold obtained through minimization technique in the objective functions P and Q show perfect agreement within error

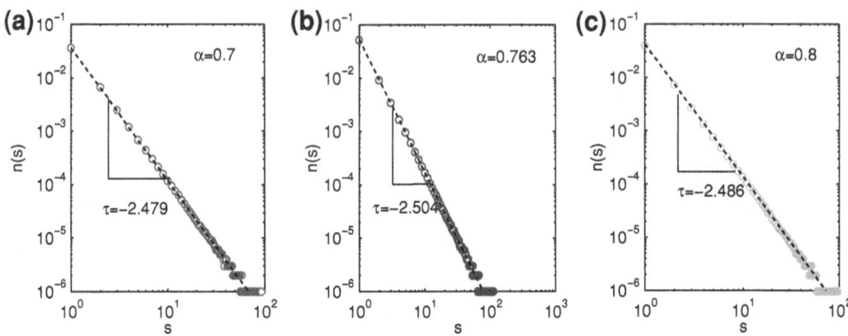

Fig. 4.6 Component size distribution $n(s)$ at the percolation threshold p_c' for $\alpha = 0.7$ (**a**), 0.763 (**b**), 0.8 (**c**), respectively

bars. Figure 4.5d–f display C_2 and χ_∞ at $p = p_c'$ as a function of system size N for $\alpha = 0.7, 0.763, 0.8$, respectively. The slope of dashed lines, which are the best fit of points, represent the exponent ratios β/ν and γ/ν in scaling relations Eqs. (4.7) and (4.8). We obtain the exponent ratios $\gamma/\nu = 0.32 \pm 0.03, 0.25 \pm 0.04, 0.23 \pm 0.05$ for $\alpha = 0.7, 0.763, 0.8$, respectively, which agree with the corresponding exponent ratios β/ν. Therefore, the validity of Eq. (4.9) has been verified. The exponent ratios $\beta/\nu, \gamma/\nu$ for $\alpha = 0.7$ are the same as what is observed for ER [17], indicating they belong to the same universality class. The measurement of exponents β, γ and ν verify the fact that the percolation transition is continuous.

Next, we numerically measure the component size distribution $n(s)$ at percolation threshold. Figure 4.6a–c show that $n(s)$ follow a power law distribution for $\alpha = 0.7, 0.763, 0.8$ with respective scaling exponent $\tau = 2.48 \pm 0.01, 2.50 \pm 0.01, 2.49 \pm 0.01$, which is further evidence for a continuous phase transition. The continuous phase transition of the second largest component is a consequence of a relatively large value of the cap k throughout the continuous growth process of $\mathbf{C_2}$. This stands in contrast to the subcritical regime for $\mathbf{C_1}$ where small values of k keep components similar in size by the mechanism of growth by overtaking.

The ratios of the scaling exponents $\beta/\nu, \gamma/\nu$, and the percolation threshold p_c' are measured by minimization technique with results shown in Table 4.2. For $\alpha \in [0.6, 0.95]$ we find strong numerical evidence that the percolation transition is indeed

Table 4.2 Summary of numerical results obtained from minimization technique

α	0.6	0.65	0.68	0.7	0.72	0.75	0.763(α_c)	0.8	0.85	0.9
β/ν	0.26(4)	0.31(4)	0.30(3)	0.31(3)	0.27(3)	0.30(3)	0.25(4)	0.24(5)	0.28(4)	0.22(5)
γ/ν	0.26(4)	0.32(4)	0.32(1)	0.32(3)	0.27(3)	0.32(3)	0.25(4)	0.23(5)	0.28(4)	0.22(5)
τ	2.50 (1)	2.48(2)	2.49(2)	2.48(1)	2.49(1)	2.49(2)	2.50(1)	2.49(1)	2.46(2)	2.48(2)
p_c'	7.595(6)	5.320(5)	4.385(5)	3.865(5)	3.455(4)	2.860(4)	2.725(3)	3.555(5)	5.215(5)	8.647(7)

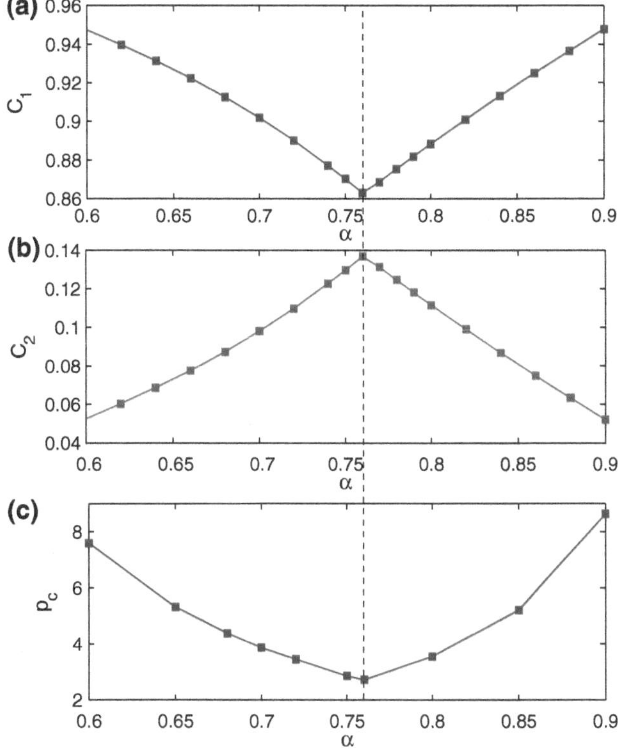

Fig. 4.7 **a** The asymptotic fraction of nodes in the largest component is plotted as a function of α. **b** The asymptotic fraction of nodes in the second largest component is plotted as a function of α. **c** The percolation threshold p_c' of the second phase transition is plotted as a function of α. The minimal values of C_1 and p_c', the maximal value of C_2 are obtained at the same critical point $\alpha_c = 0.763 \pm 0.001$, marking also the point with the minimal delay between the emergence of $\mathbf{C_1}$ and $\mathbf{C_2}$

continuous. Numerical results indicate that for $\alpha \in [0.65, 0.7]$, the BFW model and ER exhibit the same scaling exponent β, ν, γ, therefore, their percolation phase transitions belong to the same universality class. We also deduce from Table 4.2 that the minimal percolation threshold of the order parameter C_2 is obtained for $\alpha_c = 0.763 \pm 0.002$ as also shown explicitly in Fig. 4.7c. This observation shows perfect agreement with the location where $t(k_{\text{final}})$ and k_{final}/N attains minimum (Fig. 4.3a). The underlying mechanism, which accounts for the minimal percolation threshold obtained at $\alpha_c = 0.763 \pm 0.002$, is actually related with the behavior of $\mathbf{C_1}$ (or stage k) in supercritical regime. Since if $\alpha < \alpha_c$, both $k(t_c)/N$ and t_c/N decrease with α from numerical results obtained in Sect. 4.3, the emergence of the second giant component is earlier for larger α. However, if $\alpha > \alpha_c$, C_1 keeps on augmenting after t_c/N and a finite fraction of edges is added before $t(k_{\text{final}})/N$. Therefore, the competition of edges enhancing the largest component and the second largest component delays the onset of the second giant component. In particular, the emergence of the second giant component is more delayed for larger α if $\alpha > \alpha_c$. Thus, we observe the minimal percolation threshold of the order parameter C_2 is derived for $\alpha_c = 0.763 \pm 0.002$.

4.5 Summary and Discussion

In this chapter, we have analyzed the BFW model in the regime $\alpha \in [0.6, 0.95]$ where only one giant component appears in a discontinuous phase transition. The growth cessation of the largest component in the supercritical regime results in the emergence of a second giant component at some delayed time. We have carried out finite-size scaling to study the critical behavior of the second phase transition and the results establish that the transition is continuous, but that only for $\alpha \in [0.65, 0.7]$ do BFW and ER belong to the same universality class. We have also established that there exists an interesting inflection point at $\alpha_c = 0.763 \pm 0.002$ that is related to many properties of the system. As derived in Eq. (4.4), for $\alpha < \alpha_c$, $\mathbf{C_1}$ stops growing the instant it emerges, whereas for $\alpha > \alpha_c$, $\mathbf{C_1}$ grows in the supercritical regime. As shown in Fig. 4.7, at α_c the asymptotic size of $\mathbf{C_1}$ is minimized, that of $\mathbf{C_2}$ is maximized, and the delay between the emergence of $\mathbf{C_1}$ and $\mathbf{C_2}$ is minimized.

There is much recent interest in understanding the formation mechanism of multiple giant components in percolation [39, 40, 53, 54]. In Ref. [39, 40], the modified BFW model with $\alpha < 0.52$ exhibits multiple giant components which emerge in a discontinuous transition. The underlying mechanisms responsible for the formation of multiple giant components are (i) the domination of overtaking processes in the growth of $\mathbf{C_1}$, which result in the coexistence of many large components in the critical window, and (ii) the dominance of the sampling of intra-component edges in the supercritical region, which avoids merging the giant components. In Ref. [53], the so-called *multi-ER* model shows multiple giant components appear in a continuous transition. The formation of multiple giant components is attributed to the relatively low probability for merger of large components in the critical window. In this paper, the BFW model with $\alpha \in [0.6, 0.95]$ provides a novel formation mechanism resulting in multiple giant components, which is the growth cessation of $\mathbf{C_1}$ in supercritical regime resulting in the emergence of a second giant component.

The hybrid of continuous and discontinuous phase transitions observed in the studied modified BFW model is an unusual characteristic in classical statistical mechanics of critical phenomena. In a recent study, a competitive percolation model called the Nagler-Tiessen-Gutch model (NTG) [50] has been proposed in which three nodes are randomly sampled at each step and those two picked nodes are connected whose components they reside in are most similar in size. This model undergoes both continuous and discontinuous phase transitions as well. However, there is a different main feature between the NTG model and the modified BFW model with $\alpha \in [0.6, 0.9]$. In the NTG model, the double giant components would merge infinitely many times in the supercritical regime, leading to infinite discontinuous transitions of $\mathbf{C_1}$ and infinite continuous transitions of $\mathbf{C_2}$, while in the modified BFW model there are asymptotically stable double giant components that never merge.

References

1. Strogatz, S.H.: Exploring complex networks. Nature **410**, 268276 (2001)
2. Newman, M.E.J., Watts, D.J., Strogatz, S.H.: Random graph models of social networks. Proc. Natl. Acad. Sci. **99**, 2566–2572 (2002)
3. Song, C., Havlin, S., Makse, H.A.: Origins of fractality in the growth of complex networks. Nat. Phys. **2**, 275–281 (2006)
4. Buldyrev, S.V., Parshani, R., Paul, G., Stanley, H.E., Havlin, S.: Catastrophic cascade of failures in interdependent networks. Nature **464**, 1025–1028 (2010)
5. Kim, J., Krapivsky, P.L., Kahng, B., Redner, S.: Infinite-order percolation and giant fluctuations in a protein interaction network. Phys. Rev. E **66**, 055101 (2002)
6. Rozenfeld, H.D., Gallos, L.K., Makse, H.A.: Explosive percolation in the human protein homology network. Eur. Phys. J. B **75**, 305–310 (2010)
7. Moore, C., Newman, M.E.J.: Epidemics and percolation in small-world networks. Phys. Rev. E **61**, 5678–5682 (2000)
8. Serrano, M.A., Boguñá, M.: Percolation and epidemic thresholds in clustered networks. Phys. Rev. Lett. **97**, 088701 (2006)
9. Dorogovtsev, S.N., Goltsev, A.V., Mendes, J.F.F.: Critical phenomena in complex networks. Rev. Mod. Phys. **80**, 1275–1335 (2008)
10. Parshani, R., Carmi, S., Havlin, S.: Epidemic threshold for the susceptible-infectious-susceptible model on random networks. Phys. Rev. Lett. **104**, 258701 (2010)
11. Ausloos, M., Lambiotte, R.: Clusters or networks of economies? A macroeconomy study through gross domestic product. Phys. A **382**, 16–21 (2007)
12. Roca, C. P., Draief, M., Helbing, D.: Percolate or die: Multi-percolation decides the struggle between competing innovations. arXiv:1101.0775v1
13. Erdös, P., Rényi, A.: On the evolution of random graphs. Publ. Math. Inst. Hungar. Acad. Sci. **5**, 17 (1960)
14. Achlioptas, D.D., 'Souza, R.M., Spencer. J.: Explosive percolation in random networks. Science **323**, 1453–1455 (2009)
15. Cho, Y.S., Kim, J.S., Park, J., Kahng, B., Kim, D.: Percolation transitions in scale-free networks under the Achlioptas process. Phys. Rev. Lett. **103**, 135702 (2009)
16. Radicchi, F., Fortunato, S.: Explosive percolation in scale-free networks. Phys. Rev. Lett. **103**, 168701 (2009)
17. Radicchi, F., Fortunato, S.: Explosive percolation: a numerical analysis. Phys. Rev. E **81**, 036110 (2010)
18. Ziff, R.M.: Explosive growth in biased dynamic percolation on two-dimensional regular lattice networks. Phys. Rev. Lett. **103**, 045701 (2009)
19. Ziff, R.M.: Scaling behavior of explosive percolation on the square lattice. Phys. Rev. E **82**, 051105 (2010)
20. da Costa, R.A., Dorogovtsev, S.N., Goltsev, A.V., Mendes, J.F.F.: Explosive percolation transition is actually continuous. Phys. Rev. Lett. **105**, 255701 (2010)
21. Nagler, J., Levina, A., Timme, M.: Impact of single links in competitive percolation. Nat. Phys. **7**, 265–270 (2011)
22. Riordan, O., Warnke, L.: Explosive percolation is continuous. Science **333**, 322–324 (2011)
23. Grassberger, P., Christensen, C., Bizhani, G., Son, S.-W., Paczuski, M.: Explosive percolation is continuous, but with unusual finite size behavior. Phys. Rev. Lett. **106**, 225701 (2011)

24. Lee, H.K., Kim, B.J., Park, H.: Continuity of the explosive percolation transition. Phys. Rev. E **84**, 020101(R) (2011)
25. Cho, Y.S., Kahng, B., Kim, D.: Cluster aggregation model for discontinuous percolation transitions. Phys. Rev. E **81**, 030103(R) (2010)
26. Araújo, N.A.M., Herrmann, H.J.: Explosive percolation via control of the largest cluster. Phys. Rev. Lett. **105**, 035701 (2010)
27. Moreira, A.A., Oliveira, E.A., Reis, S.D.S., Herrmann, H.J., Andrade, J.S.: Hamiltonian approach for explosive percolation. Phys. Rev. E **81**, 040101(R) (2010)
28. Schrenk, K.J., Araújo, N.A.M., Herrmann, H.J.: Gaussian model of explosive percolation in three and higher dimensions. Phys. Rev. E **84**, 041136 (2011)
29. Choi, W., Yook, S.-H., Kim, Y.: Explosive site percolation with a product rule. Phys. Rev. E **84**, 020102(R) (2011)
30. Cho, Y.S., Kahng, B.: Discontinuous percolation transitions in real physical systems. Phys. Rev. E **84**, 050102(R) (2011)
31. Cho, Y. S., Kim, Y. W., Kahng, B.: Discontinuous percolation in diffusion-limited cluster aggregation. J. Stat. Mech. P10004 (2012)
32. Panagiotou, K., Spöhel, R., Steger, A., Thomas, H.: Explosive percolation in Erdös-Rényi-like random graph processes. Electron. Notes Discrete Math. **38**, 699–704 (2011)
33. Boettcher, S., Singh, V., Ziff, R.M.: Ordinary percolation with discontinuous transitions. Nat. Commun. **3**, 787 (2012)
34. Bizhani, G., Paczuski, M., Grassberger, P.: Discontinuous percolation transitions in epidemic processes, surface depinning in random media, and Hamiltonian random graphs. Phys. Rev. E **86**, 011128 (2012)
35. Cao, L., Schwarz, J.M.: Correlated percolation and tricriticality. Phys. Rev. E **86**, 061131 (2012)
36. Cho, Y.S., Kahng, B.: Suppression effect on explosive percolation. Phys. Rev. Lett. **107**, 275703 (2011)
37. Schrenk, K.J., Felder, A., Deflorin, S., Araujo, N.A.M., D'Souza, R.M., Herrmann, H.J.: Bohman-Frieze-Wormald model on the lattice, yielding a discontinuous percolation transition. Phys. Rev. E **85**, 031103 (2012)
38. Bohman, T., Frieze, A., Wormald, N.C.: Avoidance of a giant component in half the edge set of a random graph. Random Struct. Algorithms **25**, 432–449 (2004)
39. Chen, W., D'Souza, R.M.: Explosive percolation with multiple giant components. Phys. Rev. Lett. **106**, 115701 (2011)
40. Chen, W., Zheng, Z., D'Souza, R.M.: Deriving an underlying mechanism for discontinuous percolation. Europhys. Lett. **100**, 66006 (2012)
41. Bengtsson, M., Kock, S.: Cooperation and competition in relationships between competitors in business networks. J. Bus. Ind. Mark. **14**, 178–194 (1999)
42. Frank, S.A.: Repression of competition and the evolution of cooperation. Evol. Int. J. Org. Evol. **57**, 693–705 (2003)
43. Hirshleifer, J.: Competition, cooperation, and conflict in economics and biology. Am. Econ. Rev. **68** (1978)
44. Spencer, J.: The giant component: the golden anniversary. Not. AMS **57**, 720–724 (2010)
45. Ben-Naim, E., Krapivsky, P.L.: Percolation with multiple giant clusters. J. Phys. A **38**, L417–L423 (2005)
46. Asztalos, A., Toroczkai, Z.: Network discovery by generalized random walks. Europhys. Lett. **92**, 50008 (2010)
47. Anderson, R.M., May, R.M.: Infectious Diseases in Humans. Oxford University Press, Oxford (1992)
48. Manna, S.S., Chatterjee, A.: A new route to explosive percolation. Phys. A **390**, 177–182 (2011)
49. Riordan, O., Warnke, L.: Achlioptas processes are not always self-averaging. Phys. Rev. E **86**, 011129 (2012)
50. Nagler, J., Tiessen, T, Gutch, H.W.: Continuous percolation with discontinuities. Phys. Rev. X **2**, 031009 (2012)

51. Landau, D.P., Binder, K.: A Guide to Monte Carlo Simulations in Statistical Physics. Cambridge University Press, Cambridge (2000)
52. Bastas, N., Kosmidis, K., Argyrakis, P.: Explosive site percolation and finite-size hysteresis. Phys. Rev. E **84**, 066112 (2011)
53. Zhang, Y., Wei, W., Guo, B., Zhang, R., Zheng, Z.: Formation mechanism and size features of multiple giant clusters in generic percolation processes. Phys. Rev. E **86**, 051103 (2012)
54. Zhang, R., Wei, W., Guo, B., Zhang, Y., Zheng, Z.: Analysis on the evolution process of BFW-like model with discontinuous percolation of multiple giant components. Phys. A **392**, 1232–1245 (2013)

Chapter 5
Unstable Supercritical Discontinuous Percolation Transitions

5.1 Introduction

Percolation is a pervasive concept [1], which has applications in a wide variety of natural, technological and social systems [2–5], ranging from conductivity of composite materials [6, 7] and polymerization [8] to epidemic spreading [9–11] and information diffusion [12, 13]. Viewed in a network setting, once the density of edges exceeds a critical threshold, p_c, the system undergoes a sudden transition to global connectivity, where the size of the largest connected component transitions from microscopic to macroscopic in size. If, rather than edge density, we consider the impact of adding individual edges, we expect to observe the largest jump in size of the largest component at p_c.

It is well known that the classic Erdös-Rényi (ER) [14] model of percolation undergoes a continuous *second-order* phase transition during link-addition [15]. Here, one starts from a collection of N isolated nodes and edges are added uniformly at random, with the critical edge density, $p_c = 0.5$. Instead of Erdös-Rényi, we can consider *competitive percolation* processes [16]. In a competitive process, rather than a single edge, a fixed number of edges (or nodes) are chosen uniformly at random, but only the edge that best fits some specified criteria is added to the graph. Competition between edges is typically referred to as an Achlioptas Process [17].

Achlioptas Processes (AP) can exhibit a very sharp *explosive* transition, which appears discontinuous on any finite system [17]. In past years such sharp transitions have been demonstrated for scale-free networks [18–20], square lattices [21, 22], Bethe lattice [23], directed networks [24] and more realistic systems [25–27]. Although strong numerical evidence suggests that many explosive AP are discontinuous [17], more recently it has been shown that the seeming discontinuity at the percolation transition point disappears in the thermodynamic limit [16, 28–32]. However, this means that neither all AP are necessarily globally continuous nor that there are no genuine discontinuities during the first continuous emergence of a giant component. In fact, a giant connected component can emerge in a series of infinitely

W. Chen, *Explosive Percolation in Random Networks*, Springer Theses,
DOI: 10.1007/978-3-662-43739-1_5

many genuinely discontinuous jumps, and the notion that explosive percolation is always continuous [30] is thus misleading [33].

Several random network percolation models have now been identified and studied that show a single genuine discontinuous transition [29, 34–44], or even multiple discontinuities [33, 45, 46]. The mechanisms for discontinuous transitions such as dominant overtaking [16, 47], cooperative phenomena [41], and the suppression principle [48], have received considerable attention, as have criteria to discriminate between continuous and discontinuous explosive percolation transitions. A signature for a continuous percolation transition is a critical power-law component-size distribution [20, 28] and an asymptotically vanishing order parameter at the phase transition point [36].

A method to discriminate between weakly and genuinely discontinuous transitions proposed in [16] is to use the asymptotic size of the largest jump in the order parameter from the addition of a single edge. Importantly, if the largest jump of the order parameter does not vanish as the system size $N \to \infty$, the transition is necessarily discontinuous. However, whether the largest jump in the order parameter asymptotically coincides with the percolation transition point, and thus announces it, has remained largely unaddressed.

In this chapter, we introduce our work in Ref. [49]. We study whether the position of the largest jump in the order parameter asymptotically converges to the percolation transition point. To exemplify this we study the Erdös-Rényi model, Achlioptas Processes and the generalized Bohman-Frieze-Wormald model (BFW) [50]. We find that the position of the largest jump in the order parameter asymptotically converges to the percolation transition point for ER and AP with global continuity, but not necessarily for AP with discontinuities. For the BFW model, it depends on the value of the model parameter α. In BFW multiple giant components emerge at the percolation transition point, with the value of α determining the number of giants [35]. Here, we show that there are further sub-regimes of α values, with "stable" regions where the macroscopic components never merge and "unstable" regions where giants can have further merging in the supercritical regime. In stable α regions the largest jump coincides with the percolation threshold, but in unstable regions the largest jump is in the supercritical regime.

5.2 Percolation Models with Global Continuity

We study whether the position of the largest jump in the order parameter asymptotically converges to the percolation transition point for percolation models exhibiting global continuity. The best understood percolation model that shows a continuous phase transition is the Erdös-Rényi model (ER) [14]. Let N be the number of nodes, t be the number of links (i.e., edges) in system and let C_i denote the i-th largest component and C_i denote the fraction of nodes in the i-th largest component. A typical evolution of C_1, as a function of the link density $p = t/N$ (number of links per node) is shown in Fig. 5.1a. To study whether the position of the largest jump in C_1

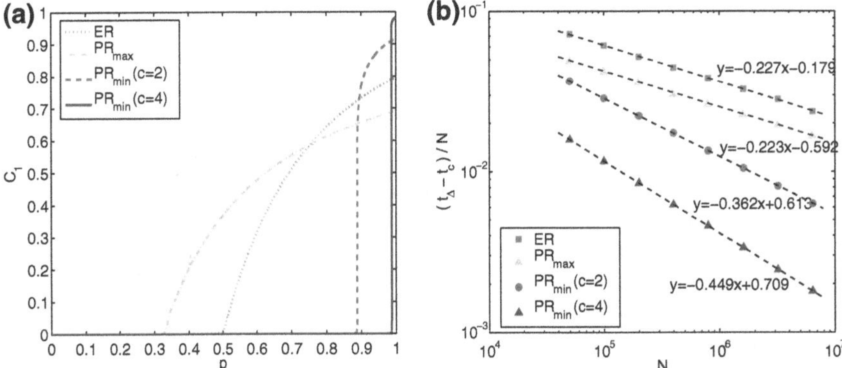

Fig. 5.1 Continuous percolation models. a A typical evolution of C_1 as a function of the link density $p = t/N$ for ER, and for the $PR_{\max}$, $PR_{\min}$ models with the number of candidate edges $c = 2$ and $c = 4$, respectively. The system size $N = 10^6$. **b** $(t_\Delta - t_c)/N$ versus number of nodes N for these models, all of which follow a power-law distribution. Each data point in (**b**) is averaged over 1,000 realizations

converges to the percolation transition point, we measure $(t_\Delta - t_c)/N$ as a function of N, where t_c denotes the minimal number of steps for C_1 to exceed $N^{1/2}$, and t_Δ is the *time* (number of steps) when the largest jump in C_1 has occurred. Figure 5.1 shows that $(t_\Delta - t_c)/N \sim N^{-0.227}$ for finite systems of very large size, which suggests that $(t_\Delta - t_c)/N$ is very likely to asymptotically converge to zero. In addition, both t_Δ/N and t_c/N are very likely to converge to the percolation transition point in the thermodynamic limit.

Next, we study two extremal AP models, specifically the $PR_{\max}$ model and the "explosive" $PR_{\min}$ model. In the $PR_{\max}$ model, two candidate links are selected randomly at each step and the link that maximizes the product of the component sizes that the ends of the link reside in is added, while the other link is discarded. In the $PR_{\min}$ model, c candidate links are selected randomly at each step and the link that minimizes the product of the component sizes that the ends of the link reside in is added, while all other links are discarded. As an example, in Fig. 5.1a we show the evolution of C_1 as a function of the link density $p = t/N$ for $PR_{\max}$ and $PR_{\min}$. Similarly to the numerical results for the ER model, we find that $(t_\Delta - t_c)/N$ follows a power-law distribution for all studied models. In particular, for finite systems of very large size, $(t_\Delta - t_c)/N \sim N^{-0.223}$ for $PR_{\max}$, $\sim N^{-0.362}$ for $PR_{\min}$ with $c = 2$, and $\sim N^{-0.449}$ for $PR_{\min}$ with $c = 4$, respectively.

This suggests that in continuous percolation, even if extremely abrupt, the largest gap asymptotically announces the percolation transition. This convergence, however, is not always guaranteed as we will show next.

5.3 Percolation Models with Discontinuities

In contrast to AP models discussed in the previous section, several random neighbor models (of the AP class) have been studied that exhibit 'staircase' discontinuities in the supercritical regime [33, 51]. Here, we focus on three types of AP models, the Devil's staircase model (DS) [33], the Nagler-Gutch model (NG) [51], and the modified ER model (mER model) [51]. The DS model is based on picking three nodes at random and forbidding the largest picked component to merge with components whose sizes are not similar, which results in a Devil's staircase with an infinite hierarchy of discontinuous jumps in C_1 [33], see Fig. 5.2a. This model has been used as a counterexample for the conclusion made in Ref. [30] that explosive percolation should be always continuous [33].

Like the DS model, the NG model and mER model are both based on 3-node rules in which the addition of links connecting two components whose sizes are similar is favored. Let p_1 denote the link density immediately after the largest jump in C_1 from the addition of a single edge, and p_2 denote the link density immediately after the second largest jump in C_1 from the addition of a single edge. Figure 5.2d shows that p_1, p_2 and $p_1 - p_2$ are very likely to asymptotically converge to some positive constant as N increases, because of the occurrence of multiple discontinuous jumps in the supercritical regime. Figure 5.2b, c show realizations of C_1 and C_2 for the NG and the mER model, in both cases featuring multiple discontinuous transitions of C_1 [51].

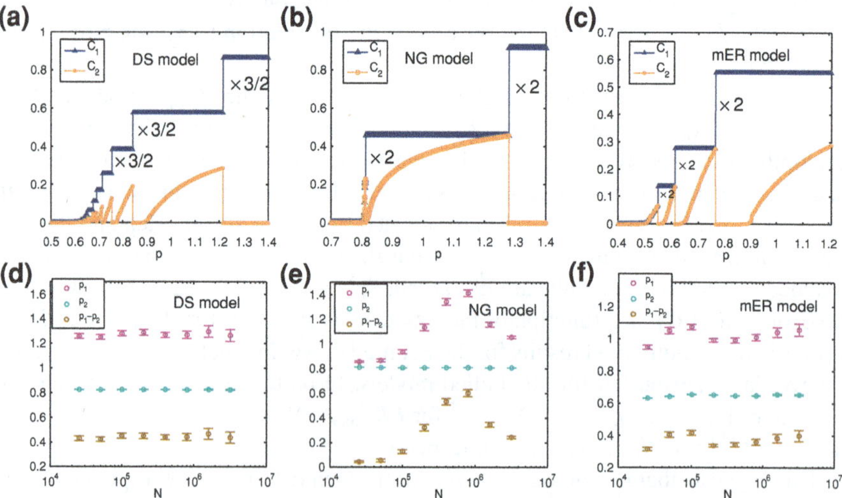

Fig. 5.2 Unstable supercritical discontinuous transitions. Typical evolutions of C_1 and C_2 versus the link density for the DS model (**a**), NG model (**b**) and mER model (**c**) with system size $N = 10^6$. The link density p_1 at which the largest jump in C_1 occurs, the link density p_2 when the second largest jump in C_1 occurs and $p_1 - p_2$ versus system size N for the DS model (**d**), NG model (**e**) and mER model (**f**). Each data point in (**d**), (**e**) and (**f**) is averaged over 100 realizations

Figure 5.2e, f show that for both the NG model and the mER model, p_1, p_2 and $p_1 - p_2$ are very likely to asymptotically converge to some non-zero constant as the system size N increases. This numerical result can be understood in the following way. From numerical observations, the size of the third largest component immediately before p_2 in the supercritical regime is at most $\mathcal{O}(\log N)$ for the DS model, the NG model and mER model. In addition, for all these models it has been analytically demonstrated that immediately before p_2 both the size of the largest and second largest component are of order $\mathcal{O}(N)$, where for the DS model it can be shown that $C_2 = \frac{1}{2}C_1$, while for the NG model and the mER model $C_2 = C_1$ [33, 51]. As a result, for these models once the largest and second largest component merge together inducing a discontinuous jump, the size of the second largest component drops to $\mathcal{O}(\log N)$. Thus $\mathcal{O}(N)$ links are required before at p_1 the second largest component grows again to size $\frac{1}{2}C_1$ (for the DS model), or to size C_1 for the NG model and the mER model, respectively. Thus, $p_1 - p_2$ is necessarily extensive and asymptotically converges to some positive non-zero constant.

This demonstrates the occurrence of multiple discontinuous transitions, including the transition with the largest discontinuity, in the supercritical regime and not at the percolation critical point as in traditional percolation.

In general, we use the term stable coexistence when all giant components emerging at the percolation transition point persist and remain separate throughout the supercritical regime. We use the term unstable coexistence when at least two giant components emerging at the percolation transition point merge together at some point in the supercritical regime. We find that all the models studied in this section display unstable supercritical discontinuous transitions, which is a novel and unexpected feature in percolation. The model we study next shows even a quantitatively richer behavior, with some regions of stable coexistence and other regions of unstable coexistence.

5.4 Bohman-Frieze-Wormald Model

In this section, we investigate the Bohman-Frieze-Wormald (BFW) Model. We first briefly illustrate the process of the BFW model. The system is initialized with N isolated nodes. A cap, k, on the maximum allowed component size is initially set to $k = 2$ and increased in stages as follows. Links are sampled one-at-a-time, uniformly at random from the complete graph generated by the N nodes. If a link would lead to formation of a component of size less than or equal to k it is accepted. Otherwise, the link is rejected provided that the fraction of accepted links remains greater than or equal to a function $g(k) = \alpha + (2k)^{-1/2}$, where α is a tunable parameter (In the original BFW model, Bohman, Frieze, and Wormald only studied the case of $\alpha = 1/2$). Finally, if the fraction of accepted links would drop below $g(k)$, the cap is augmented to $k + 1$, and the impact of adding the link reevaluated against the new values, $k + 1$ and $g(k + 1)$. This final step of augmenting the cap and reevaluating the impact is iterated until either k increases sufficiently to accept the link or $g(k)$

decreases sufficiently that the link can be rejected. Asymptotically, the fraction of accepted links is $\lim_{k\to\infty} g(k) = \alpha$. (The details of BFW model can be found in the appendix.)

The original BFW model ($\alpha = 1/2$) was first introduced to demonstrate that the emergence of a giant component can be largely suppressed [50]. In particular, it has been established that under the BFW evolution, if $m = 0.96689N$ links out of $2m$ sequentially sampled random links have been added to a graph, a giant component does not exist [50]. More recently, the nature of the BFW transition was investigated. It was established that multiple giant components appear in a discontinuous percolation transition for the BFW model [35]. In the asymptotic time limit of the original BFW model, one-half of all links that are sampled must be added. Generalizing the BFW model by allowing the asymptotic fraction of accepted links to be a parameter α, then the number of giant components can be tuned by adjusting the value of α [35].

For this model, we investigate whether discontinuous jumps of the order parameter occur at the percolation transition point or in the supercritical regime. First, we study the evolution of the size of the four largest components, C_i with $i = 1, 2, 3, 4$ for the BFW model with $\alpha \in (0, 1]$, as a function of the link density p.

In Fig. 5.3a we show the discontinuous emergence of a unique giant component in a sharp transition to global connectivity, for $\alpha = 0.6$. For other values of α, multiple giant components emerge simultaneously in a sharp transition. See, for instance, Fig. 5.3c for $\alpha = 0.5$ with two giant components, or Fig. 5.3e for $\alpha = 0.3$ with three giant components. Thus, for certain values of α, there exists a unique transition to global connectivity with multiple giant components.

Yet, Fig. 5.3b, d, and f show there is another type of behavior possible, with an additional transition in the supercritical regime where giant components merge. This suggests the existence of an instability of giant components. Next, we perform numerical simulations to characterize the occurrence of discontinuities and multiplicities during the discontinuous transitions.

As in the previous section, we denote the link density p_1 as the position immediately after the largest jump in C_1 from the addition of a single edge, and p_2 the position immediately after the second largest jump in C_1 from the addition of a single edge. In addition, let p_3 denote the minimal position at which the largest component contains at least $N^{1/2}$ nodes. From numerical results in Fig. 5.3, we find that $p_1 \geq p_2 \geq p_3$.

Let us focus initially on the region with stable coexisting giant components. For $\alpha = 0.6, 0.5, 0.3$, Fig. 5.4a, c, and e show that the size of the largest component at p_1, denoted as $C_1(p_1)$, is almost independent of the system size N, and converges to some positive constant asymptotically. On the other hand, the gap between p_1 and p_3, i.e., $\Delta p = p_1 - p_3$, scales as a negative power of N, and thus is very likely to decrease to zero as $N \to \infty$, see Fig. 5.4. This suggests that once the number of nodes in the largest component increases from $\mathcal{O}(N^{1/2})$ to $\mathcal{O}(N)$, the augmented link density converges to zero as $N \to \infty$, indicating the percolation process undergoes a unique discontinuous transition in the thermodynamic limit.

We further find that at p_1, for $\alpha = 0.5$, C_2 is very likely to converge to some positive constant as well, see Fig. 5.4c, and for $\alpha = 0.3$, C_1, C_2, and C_3 are all very likely to converge to some positive nonzero constant, respectively, see Fig. 5.4e.

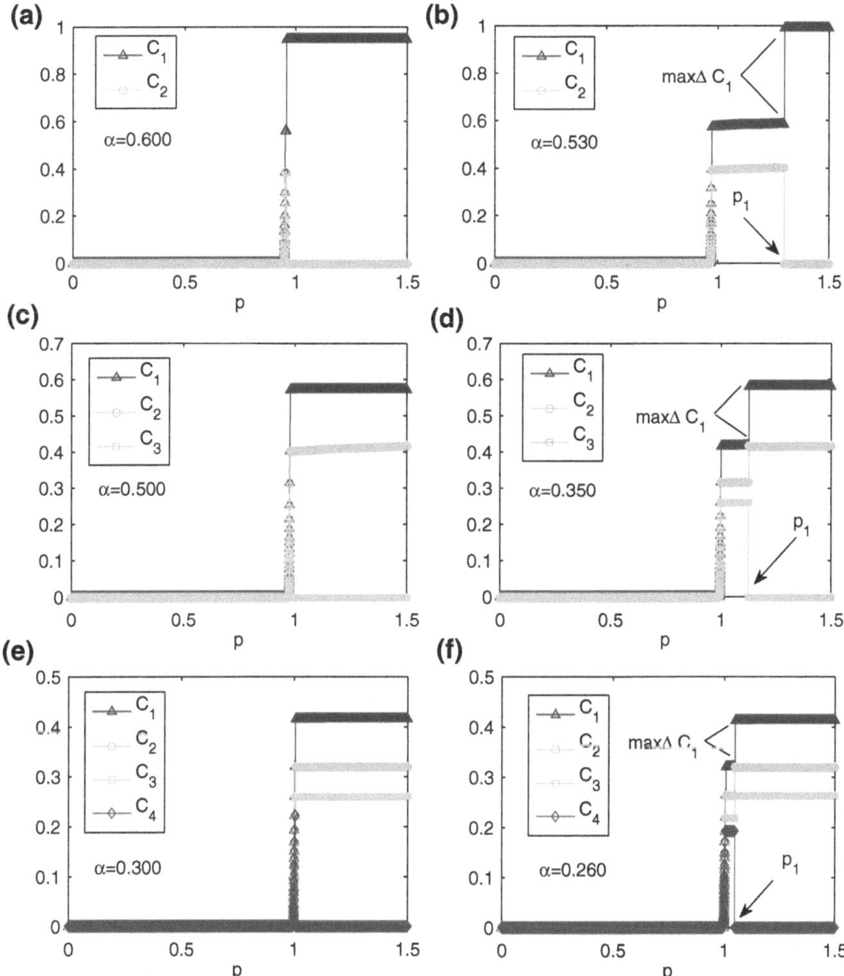

Fig. 5.3 Critical and unstable supercritical discontinuous transitions. **a** For $\alpha = 0.600$, one giant component emerges in a phase transition. **b** For $\alpha = 0.530$, C_1, C_2 versus the link density, showing that two giant components emerge simultaneously in the first phase transition. They are, however, unstable as they merge at a second transition. **c** For $\alpha = 0.500$, C_1, C_2, C_3 versus the link density, showing that two giant components emerge simultaneously. **d** For $\alpha = 0.350$, C_1, C_2, C_3 versus the link density, showing that three giant components emerge simultaneously in the first phase transition. This configuration is unstable as in a second transition the second largest and the third largest components merge. **e** For $\alpha = 0.300$, C_1, C_2, C_3, C_4 versus link density, showing the simultaneous emergence of three giant components. **f** For $\alpha = 0.260$, C_1, C_2, C_3, C_4 versus density of links, showing that four giant components emerge simultaneously in the first phase transition. Two smallest macroscopic components C_3 and C_4 merge together and overtake C_1, the other macroscopic components are stable in the remaining process, $C_3 + C_4 \rightarrow C_1, C_1 \rightarrow C_2, C_2 \rightarrow C_3$. System size is $N = 10^6$ for all simulations

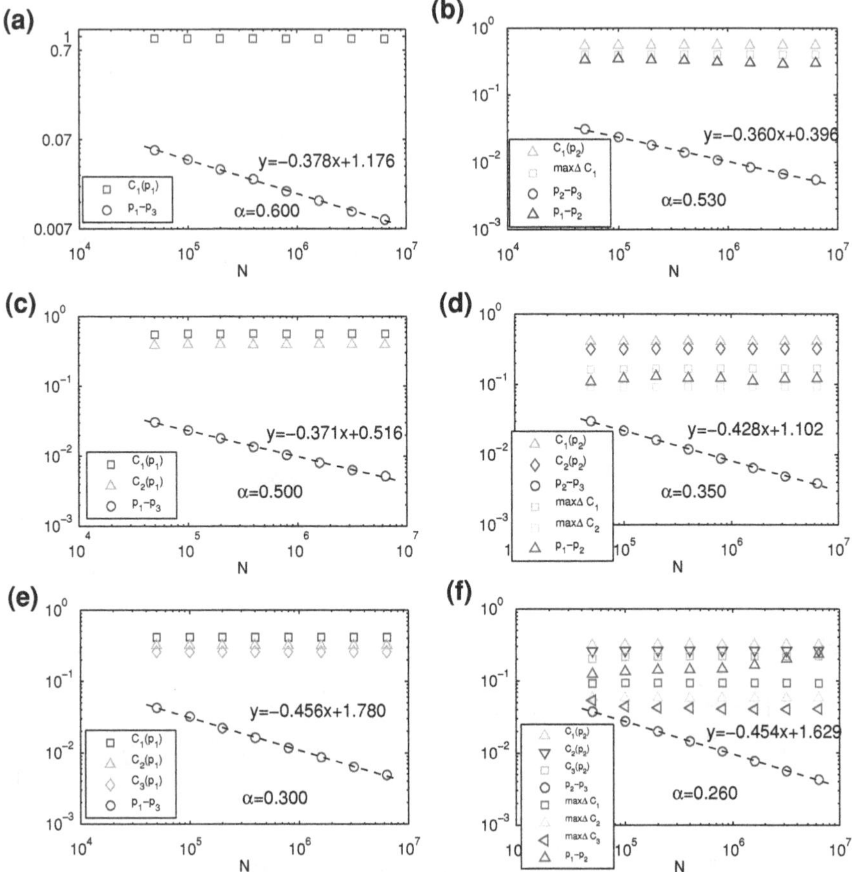

Fig. 5.4 Scaling of critical and unstable supercritical discontinuous transitions. **a** For $\alpha = 0.600$, C_1 at $p_1, p_1 - p_3$ versus system size. **b** For $\alpha = 0.530$, C_1 at p_2, the largest jump in C_1, $p_2 - p_3$, $p_1 - p_2$ versus system size. **c** For $\alpha = 0.500$, C_1, C_2 at p_1, $p_1 - p_3$ versus system size. **d** For $\alpha = 0.350$, C_1, C_2 at p_2, the largest jump in $C_1, C_2, p_2 - p_3, p_1 - p_2$ versus system size. **e** For $\alpha = 0.300$, C_1, C_2, C_3 at $p_1, p_1 - p_3$ versus system size. **f** For $\alpha = 0.260$, C_1, C_2, C_3 at p_2, the largest jump in $C_1, C_2, C_3, p_2 - p_3, p_1 - p_2$, versus system size. Each data point is averaged over 1,000 realizations

This indicates that the multiple giant components appear simultaneously in a unique discontinuous transition, consistent with the theory and observations put forth in [35, 47].

Let us now focus on the regime with unstable coexistence of multiple giant components. For α in the unstable regime, the size of the largest component at p_2, is almost independent of the system size N and converges to some positive constant asymptotically. For $\alpha = 0.53, 0.35, 0.26$, the size of the two, three, and four largest components at p_2 are very likely to converge to some positive constant asymptotically. Yet, we find that $p_2 - p_3$ decays as a power law in N, see Fig. 5.4b, d, and f.

This suggests the percolation transition is discontinuous at p_2 with multiple giant components emerging simultaneously.

In addition, we observe that the size of the largest jump of the largest component at p_1, denoted by max ΔC_1, is independent of system size N, and is very likely to converge to some positive constant asymptotically (see Fig. 5.4b, d, and f). This suggests the occurrence of a second discontinuous transition at p_1 in the supercritical regime. We find that the second transition results from the merging of the two smallest giant components that emerge at the first (i.e., percolation) transition. This mechanism can be seen clearly from the case of $\alpha = 0.26$ in Fig. 5.3f, where four giant components emerge at the first transition, while at the second transition, C_3 and C_4 merge together and overtake C_1 in size. Thus, C_1, C_2, C_3 all get a sudden jump in size, but C_4 breaks down. This overtaking mechanism dominates the growth of the largest component in the BFW model, which has been proven to be a key mechanism leading to discontinuous percolation transitions [16, 33, 47].

To test if the transition points p_1 and p_2 in the unstable regime are still distinct in the thermodynamic limit, we next perform a scaling analysis. For the values of $\alpha = 0.53, 0.35, 0.26$, which are in the unstable region, we find that $p_1 - p_2$ is very likely to converge to a nonzero constant, see Fig. 5.4. This suggests, indeed, the distinctness of the two transition points.

Taken together, we have identified and studied two parameter regimes in the BFM model, (i) the stable regime of a unique discontinuous transition where one or more giant components emerge and coexist throughout the supercritical regime and (ii) the unstable regime of multiple discontinuous transitions where multiple giant components emerge but the two smallest ones merge at a well-defined transition point in the supercritical regime.

However, the characterization remains incomplete as so far we have only studied three instances for each parameter regime. Our next aim is to establish a phase diagram by continuously tuning the parameter α.

5.5 Phase Diagram of the BFW Model

We first investigate the behavior in the regime $\alpha > \alpha_1$ with $\alpha_1 = 0.511 \pm 0.003$, where only one giant component asymptotically remains in the system. For values $\alpha < \alpha_1$, two giant components asymptotically remain [35].

Since in a stable regime of α all giant components that have emerged remain separate throughout the supercritical regime, a stable regime is characterized by a unique transition of the largest component. In contrast, an unstable regime is characterized by one (or more) discontinuous transitions of C_1 in the supercritical regime by aggregation of two (or more) giant components emerging at percolation transition point.

We find that the model undergoes two distinct discontinuous phase transitions for $\alpha \in (\alpha_1, \alpha_1')$, referred to as the unstable regime, but undergoes a unique discontinuous phase transition for $\alpha \in (\alpha_1', 1)$, referred to as the stable regime, where $\alpha_1' = 0.551\pm$

Table 5.1 Summary of bifurcation points α_i and α'_i

i	α_i	α'_i
1	0.511 ± 0.003	0.551 ± 0.001
2	0.343 ± 0.001	0.352 ± 0.001
3	0.259 ± 0.001	0.261 ± 0.001
4	0.206 ± 0.001	0.208 ± 0.001

0.001. Previous work has established an infinite number bifurcation points α_i, at which the number of stable giant components that asymptotically remain in the system changes from i to $i+1$, $i \geq 1$ (see Table 5.1) [35]. Here, we numerically expand the analysis including transitions of the second largest component, the third largest component, and the forth largest component as well. These transitions lead to multiple discontinuous transitions of the largest component and a hierarchy of stable and unstable regimes. In particular, for $\alpha \in (\alpha'_i, \alpha_{i-1})$, $i \geq 2$, we identify a stable regime, where in a unique discontinuous transition the i largest macroscopic components, $C_1, C_2, ..., C_i$, simultaneously emerge, and for $\alpha \in (\alpha_i, \alpha'_i)$ an unstable regime with two distinct discontinuous transitions, for $i \geq 1$. We find the numerical values $\alpha'_2 = 0.352 \pm 0.001, \alpha'_3 = 0.261 \pm 0.001, \alpha'_4 = 0.208 \pm 0.001...$, see Table 5.1.

In Fig. 5.5 we demonstrate the phase diagram of the BFW model. We show the number of giant components that emerge at the first transition (percolation transition point) versus α, with alternating stable and unstable regions. Since the number of giant components that asymptotically remain in the system increases as α decreases, there exist infinitely many bifurcation points α'_i, $i \geq 1$, which separate stable from unstable regimes.

5.6 Summary and Discussions

For various models with continuous, discontinuous, and multiple-discontinuous percolation transitions, we have investigated whether the location of the largest jump in the order parameter asymptotically converges to the percolation transition point marking the onset of global connectivity.

For globally continuous transitions, including certain Achlioptas processes, the location of the largest jump in the order parameter asymptotically converges to the percolation transition point. In contrast, Achlioptas processes with discontinuities exhibit a "staircase" of discontinuities in the supercritical region and the location of the largest jump is at an edge density well above the percolation transition point. Finally, the BFW model exhibits a rich supercritical behavior that is dependent on the model parameter α, as exemplified by the phase diagram, Fig. 5.5, together with analytics suggesting an infinite hierarchy of regimes of alternating stability type. Whether the percolation transition is asymptotically announced by the largest gap in the size of the largest component depends on the parameter. In the stable regime

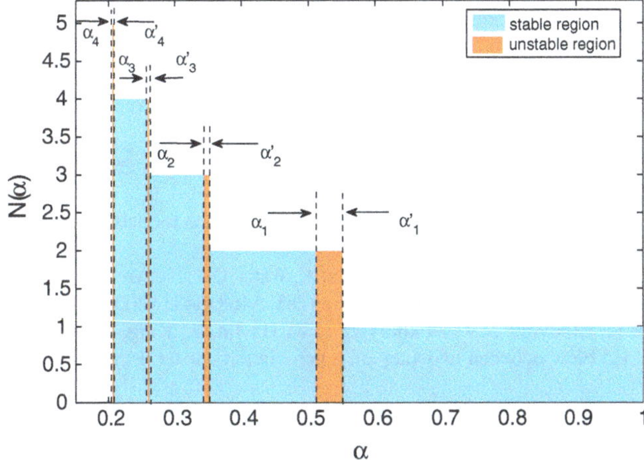

Fig. 5.5 Phase diagram for the BFW model. The number of giant components that appear in the first discontinuous phase transition (percolation transition point) in dependence on α. In the stable region, the giant components remain distinct in the supercritical regime, while in the unstable region the two smallest giant components merge in the supercritical regime at a well-defined transition point

macroscopic components robustly coexist, displaying the largest jump of the order parameter at the percolation transition point. In the unstable region the coexistence of all macroscopic components that have emerged occurs in a finite sized window only, leading to multiple discontinuous transitions. Macroscopic components that emerge at the percolation transition are thus not necessarily stable in the thermodynamic limit.

For AP models with discontinuities and for the BFW model, multiple discontinuous transitions are a consequence of the occurrence of extended periods in time well in the supercritical regime where macroscopic components cannot merge. Mechanisms implying such periods are yet to be discovered.

Multiple transitions have been studied in a wide variety of fields, such as geophysics [52], liquid crystals [53], classical thermodynamics and solid state physics [54]. However, in random network percolation multiple transitions are poorly understood. It would thus be interesting to identify the sufficient conditions for these [33, 55]. This numerical work represents a step toward this direction.

In short, we have investigated unstable discontinuous transitions in percolation. Seemingly and genuinely discontinuous percolation transition can involve a rich behavior in supercriticality, a regime that deserves attention for percolating systems with substantial delays [55]. However, our results obtained in this paper are based on numerical simulations on finite systems for practical purposes ($N < 10^7$) since real-world networks such as social networks or the Internet, are by definition of finite size, so the implications of asymptotic theory ($N \to \infty$) may not apply. It would thus be interesting to understand whether our results hold for infinite systems through a rigorous manner, which remains a challenge to be explored.

References

1. Stauffer, D., Aharony, A.: Introduction to percolation theory. Taylor & Francis, London (1994)
2. Drossel, B., Schwabl, F.: Self-organized critical forest-fire model. Phys. Rev. Lett. **69**, 1629–1632 (1992)
3. Buldyrev, S.V., Parshani, R., Paul, G., Stanley, H.E., Havlin, S.: Catastrophic cascade of failures in interdependent networks. Nature **464**, 1025–1028 (2010)
4. Newman, M.E.J., Watts, D.J., Strogatz, S.H.: Random graph models of social networks. Proc. Natl. Acad. Sci. **99**, 2566–2572 (2002)
5. Callaway, D.S., Newman, M.E.J., Strogatz, S.H., Watts, D.J.: Network robustness and fragility: percolation on random graphs. Phys. Rev. Lett. **85**, 5468–5471 (2000)
6. Andrade, J.S., Buldyrev, S.V., Dokholyan, N.V., Havlin, S., King, P.R., Lee, Y.K., Paul, G., Stanley, H.E.: Flow between two sites on a percolation cluster. Phys. Rev. E **62**, 8270–8281 (2000)
7. Sahimi, M.: Applications of Percolation Theory. Taylor & Francis, London (1994)
8. Ziff, R.M., Hendriks, E.M., Ernst, M.H.: Critical properties for gelation: a kinetic approach. Phys. Rev. Lett. **49**, 593 (1982)
9. Moore, C., Newman, M.E.J.: Epidemics and percolation in small-world networks. Phys. Rev. E **61**, 5678–5682 (2000)
10. Pastor-Satorras, R., Vespignani, A.: Epidemic spreading in scale-free networks. Phys. Rev. Lett. **86**, 3200–3203 (2001)
11. Anderson, R.M., May, R.M.: Infectious Diseases of Humans: Dynamics and Control. Oxford University Press, Oxford (1991)
12. Strang, D., Soule, S.: Diffusion in organizations and social movements: from hybrid corn to poison pills. Ann. Rev. Sociol. **24**, 265290 (1998)
13. Lazarsfeld, P.F., Berelson, B., Gaudet, H.: The Peoples Choice: How the Voter Makes up his Mind in a Presidential Campaign. Columbia University Press, New York (1944)
14. Erdös, P., Rényi, A.: On the evolution of random graphs. Publ. Math. Inst. Hungar. Acad. Sci. **5**, 17 (1960)
15. Bollobás, B.: Random Graphs. Cambridge University Press, Cambridge (2001)
16. Nagler, J., Levina, A., Timme, M.: Impact of single links in competitive percolation. Nat. Phys. **7**, 265–270 (2011)
17. Achlioptas, D.D., 'Souza, R.M., Spencer. J.: Explosive percolation in random networks. Science **323**, 1453–1455 (2009)
18. Cho, Y.S., Kim, J.S., Park, J., Kahng, B, Kim, D.: Percolation transitions in scale-free networks under the Achlioptas process. Phys. Rev. Lett. **103**, 135702 (2009)
19. Radicchi, F., Fortunato, S.: Explosive percolation in scale-free networks. Phys. Rev. Lett. **103**, 168701 (2009)
20. Radicchi, F., Fortunato, S.: Explosive percolation: a numerical analysis. Phys. Rev. E **81**, 036110 (2010)
21. Ziff, R.M.: Explosive growth in biased dynamic percolation on two-dimensional regular lattice networks. Phys. Rev. Lett. **103**, 045701 (2009)
22. Ziff, R.M.: Scaling behavior of explosive percolation on the square lattice. Phys. Rev. E **82**, 051105 (2010)
23. Chae, H., Yook, S.-H., Kim, Y.: Explosive percolation on the Bethe lattice. Phys. Rev. E **85**, 051118 (2012)
24. Squires, S., Sytwu, K., Alcala, D., Antonsen, T.M., Ott, E., Girvan, M.: Not with a bang: weakly explosive percolation in directed networks. Phys. Rev. E **87**, 052127 (2013)
25. Kim, Y., Yun, Y.-K., Yook, S.-H.: Explosive percolation in a nanotube-based system. Phys. Rev. E **82**, 061105 (2010)
26. Rozenfeld, H.D., Gallos, L.K., Makse, H.A.: Explosive percolation in the human protein homology network. Eur. Phys. J. B **75**, 305–310 (2010)

27. Pan, R.K., Kivelä, M., Saramäki, J., Kaski, K., Kertész, J.: Using explosive percolation in analysis of real-world networks. Phys. Rev. E **83**, 046112 (2011)
28. da Costa, R.A., Dorogovtsev, S.N., Goltsev, A.V., Mendes, J.F.F.: Explosive percolation transition is actually continuous. Phys. Rev. Lett. **105**, 255701 (2010)
29. Manna, S.S., Chatterjee, A.: A new route to explosive percolation. Phys. A **390**, 177–182 (2011)
30. Riordan, O., Warnke, L.: Explosive percolation is continuous. Science **333**, 322–324 (2011)
31. Grassberger, P., Christensen, C., Bizhani, G., Son, S.-W., Paczuski, M.: Explosive percolation is continuous, but with unusual finite size behavior. Phys. Rev. Lett. **106**, 225701 (2011)
32. Lee, H.K., Kim, B.J., Park, H.: Continuity of the explosive percolation transition. Phys. Rev. E **84**, 020101(R) (2011)
33. Nagler, J., Tiessen, T, Gutch, H.W.: Continuous percolation with discontinuities. Phys. Rev. X **2**, 031009 (2012)
34. Cho, Y.S., Kahng, B., Kim, D.: Cluster aggregation model for discontinuous percolation transitions. Phys. Rev. E **81**, 030103(R) (2010)
35. Chen, W., D'Souza, R.M.: Explosive percolation with multiple giant components. Phys. Rev. Lett. **106**, 115701 (2011)
36. Araújo, N.A.M., Herrmann, H.J.: Explosive percolation via control of the largest cluster. Phys. Rev. Lett. **105**, 035701 (2010)
37. Choi, W., Yook, S.-H., Kim, Y.: Explosive site percolation with a product rule. Phys. Rev. E **84**, 020102(R) (2011)
38. Cho, Y.S., Kahng, B.: Discontinuous percolation transitions in real physical systems. Phys. Rev. E **84**, 050102(R) (2011)
39. Panagiotou, K., Spöhel, R., Steger, A., Thomas, H.: Explosive percolation in Erdös-Rényi-like random graph processes. Electron. Notes Discrete Math. **38**, 699–704 (2011)
40. Boettcher, S., Singh, V., Ziff, R.M.: Ordinary percolation with discontinuous transitions. Nat. Commun. **3**, 787 (2012)
41. Bizhani, G., Paczuski, M., Grassberger, P.: Discontinuous percolation transitions in epidemic processes, surface depinning in random media, and Hamiltonian random graphs. Phys. Rev. E **86**, 011128 (2012)
42. Cao, L., Schwarz, J.M.: Correlated percolation and tricriticality. Phys. Rev. E **86**, 061131 (2012)
43. Chen, W., Nagler, J., Cheng, X., Jin, X., Shen, H., Zheng, Z., D'Souza, R.M.: Phase transitions in supercritical explosive percolation. Phys. Rev. E **87**, 052130 (2013)
44. Cho, Y.S., Hwang, S., Herrmann, H.J., Kahng, B.: Avoiding a spanning cluster in percolation models. Science **339**, 1185–1187 (2013)
45. Schrenk, K.J., Araújo, N.A.M., Herrmann, H.J.: Gaussian model of explosive percolation in three and higher dimensions. Phys. Rev. E **84**, 041136 (2011)
46. Schrenk, K.J., Felder, A., Deflorin, S., Araújo, N.A.M., D'Souza, R.M., Herrmann, H.J.: Bohman-Frieze-Wormald model on the lattice, yielding a discontinuous percolation transition. Phys. Rev. E **85**, 031103 (2012)
47. Chen, W., Zheng, Z., D'Souza, R.M.: Deriving an underlying mechanism for discontinuous percolation. Europhys. Lett. **100**, 66006 (2012)
48. Cho, Y.S., Kahng, B.: Suppression effect on explosive percolation. Phys. Rev. Lett. **107**, 275703 (2011)
49. Chen, W., Cheng, X., Zheng, Z., Chung, N.N., D'Souza, R.M., Nagler, J.: Unstable supercritical discontinuous percolation transitions. Phys. Rev. E **88**, 042152 (2013)
50. Bohman, T., Frieze, A., Wormald, N.C.: Avoidance of a giant component in half the edge set of a random graph. Random Struct. Algorithms **25**, 432–449 (2004)
51. Riordan, O., Warnke, L.: Achlioptas processes are not always self-averaging. Phys. Rev. E **86**, 011129 (2012)
52. Ramirez, A.P., Shastry, B.S., Hayashi, A., Krajewski, J.J., Huse, D.A., Cava, R.J.: Multiple field-induced phase transitions in the geometrically frustrated dipolar magnet: Gd2Ti2O7. Phys. Rev. Lett. **89**, 067202 (2002)
53. Strzelecka, T.E., Davidson, M.W., Rill, R.L.: Multiple liquid crystal phases of DNA at high concentrations. Nature **331**, 457 (1988)

54. Armstrong, J.N., Felske, J.D., Chopra, H.D.: Multiple phase transitions found in a magnetic Heusler Alloy and thermodynamics of their magnetic internal energy. Phys. Rev. B **81**, 174405 (2010)
55. Schröder, M., Ebrahimnazhad Rahbari, S.H., Nagler, J.: Crackling noise in fractional percolation. Nat. Commun. **4**, 2222 (2013)

Appendix A
Algorithm of Percolation Models

We state the algorithm of the percolation models studied in this book in detail, which are the Devil's staircase model (DS), the Nagler-Gutch model (NG), the modified Erdös-Rényi model, and the Bohman-Frieze-Wormald model.

A.1 Devil's Staircase Model

Start with an empty graph with N isolated nodes. At each step, three nodes v_1, v_2, v_3 are randomly selected from N nodes, and let s_1, s_2, s_3 denote the sizes of components (not necessarily distinct) in which they reside. Consider $\Delta_{i,j} = |s_i - s_j|$ with $1 \leq i < j \leq 3$ and connect v_i, v_j for which $\Delta_{i,j}$ is minimal.

A.2 Nagler-Gutch Model

Start with an empty graph with N isolated nodes. At each step, three nodes v_1, v_2, v_3 are randomly selected from N nodes, and let s_1, s_2, s_3 denote the sizes of components (not necessarily distinct) in which they reside. Let s_i denote the size of the component containing v_i. If all three component sizes s_i are equal, add the edge connecting $v_1 v_2$. If exactly two component sizes s_i are equal, connect the corresponding nodes by a link. Otherwise (if all s_i are different), link the nodes in the two smallest components. This model is a modification of the "explosive" triangle rule introduced in [1], which in contrast exhibits a steep but continuous transition.

W. Chen, *Explosive Percolation in Random Networks*, Springer Theses,
DOI: 10.1007/978-3-662-43739-1

A.3 Modified Erdös-Rényi Model

Let L_1 and L_2 denote the sizes of the two largest components of the evolving graph. The mER model proceeds as follows. If the two largest components in the current graph have the same size ($L_1 = L_2$), add the link connecting $v_1 v_2$. When $L_1 > L_2$, if at least two s_i are equal to L_1, connect two corresponding nodes, otherwise connect two nodes in components of size smaller than L_1.

A.4 Bohman-Frieze-Wormald Model

Let k denote the cap on the maximum allowed component size, which is initially set to $k = 2$, N denotes the number of nodes in the graph, u denotes the total number of links sampled, A denotes the set of accepted links (initially $A = \emptyset$), and $t = |A|$ denotes the number of accepted links. At each step u, a link e_u is sampled uniformly at random from the complete graph generated by the N nodes and evaluated by the following algorithm. This algorithm iterates until a specified number of links, ωN, have been accepted:

Repeat until $u = \omega N$
{
  Set $l =$ **size of largest component in** $A \cup \{e_u\}$
  if $(l \leq k)$ {
    $A \leftarrow A \cup \{e_u\}$
    $u \leftarrow u + 1$ }
  else if $(t/u < g(k))$ { $k \leftarrow k + 1$ }
  else { $u \leftarrow u + 1$ }
}

where $g(k) = \alpha + (1/2k)^{\beta}$ with $\alpha \in [0, 1], \beta \in [0, \infty]$. Thus, α denotes the asymptotic fraction of accepted links over totally sampled links while β denotes the rate of $g(k)$ converging to its limiting value 1/2. In the original BFW model, $\alpha = 1/2, \beta = 1/2$. We generalize the BFW model by tuning the parameter α.

Reference

1. D'Souza, R.M., Mitzenmacher, M.: Local cluster aggregation models of explosive percolation. Phys. Rev. Lett. **104** , 195702 (2010)

Index

W. Chen, *Explosive Percolation in Random Networks*, Springer Theses,
DOI: 10.1007/978-3-662-43739-1

Zeitfracht Medien GmbH
Ferdinand-Jühlke-Straße 7
99095 Erfurt, Deutschland
produktsicherheit@kolibri360.de